Thinking Matters

*Critical Thinking as
Creative Problem Solving*

Thinking Matters

*Critical Thinking as
Creative Problem Solving*

Gary R. Mar
Stony Brook University, USA

NEW JERSEY · LONDON · SINGAPORE · BEIJING · SHANGHAI · HONG KONG · TAIPEI · CHENNAI · TOKYO

Published by

World Scientific Publishing Co. Pte. Ltd.

5 Toh Tuck Link, Singapore 596224

USA office: 27 Warren Street, Suite 401-402, Hackensack, NJ 07601

UK office: 57 Shelton Street, Covent Garden, London WC2H 9HE

Library of Congress Control Number: 2021029593

British Library Cataloguing-in-Publication Data
A catalogue record for this book is available from the British Library.

The cover illustration is a fractal image of a dualist form of semantic paradox based on the "Chaotic Liar" first published in Mar and Grim [1991] "Pattern and Chaos: New Images in the Semantics of Paradox," Noûs, XXV (Dec.), 659-93, discussed in Ian Stewart's [1992] "A Partly True Story", a descendant of Martin Gardner's legendary column in Scientific American, and collected with related research in Grim, Mar and St. Denis in The Philosophical. Computer: Exploratory Essays in Philosophical Computer Modeling [MIT 1998].

THINKING MATTERS
Critical Thinking as Creative Problem Solving

Copyright © 2021 by World Scientific Publishing Co. Pte. Ltd.

All rights reserved. This book, or parts thereof, may not be reproduced in any form or by any means, electronic or mechanical, including photocopying, recording or any information storage and retrieval system now known or to be invented, without written permission from the publisher.

For photocopying of material in this volume, please pay a copying fee through the Copyright Clearance Center, Inc., 222 Rosewood Drive, Danvers, MA 01923, USA. In this case permission to photocopy is not required from the publisher.

ISBN 978-981-121-684-8 (hardcover)
ISBN 978-981-121-624-4 (paperback)
ISBN 978-981-121-625-1 (ebook for institutions)
ISBN 978-981-121-626-8 (ebook for individuals)

For any available supplementary material, please visit
https://www.worldscientific.com/worldscibooks/10.1142/11712#t=suppl

Contents

Preface — vii

1 The Thinking Reed — 1
- The Wonder of Thinking 1
- Thought & Creativity 3
- Creative Thinking & Logical Inference 6
- Summary of Concepts 9
- Exercises 10

2 "Eureka!" Problem Solving Heuristics — 19
- Characterizing Problems 19
- Isolate the Real Problem 22
- Working Forward and Backward 27
- Changing Your Conceptual Mode 29
- Divide and Conquer 32
- Let the Problem Incubate 38
- Summary of Concepts 40
- Exercises 40

3 Bridges to Problem Solving — 51
- The Puzzle of the Königsberg Bridges 51
- Specializing & Generalizing 53
- Euler's Theorem 55
- Some Advantages of Modeling 56
- Elaborating a Solution 61
- Summary of Concepts 63
- Exercises 64

4 Puzzles, Paradoxes & Previews — 69
- Matrix Logic 69
- Liars, Truth-Tellers & Truth Tables 73

Inference Rules & Fallacies 83
Logical Deductions & Tree Diagrams 90
Syllogisms & Venn Diagrams 97
Rhetoric & Persuasion 102
Science or Pseudoscience? 107
Probability, Statistics & Decision Making 110
A Parting Paradox 116
Summary of Concepts 117
Exercises 120

5 Computational Magic · 137
A "Digital Computer" 137
Dialing Up An Algorithm 138
Programming for People 140
31 Flavours of Freedom 141
The Josephus Problem 143
Alan Turing: the Enigma 147
Summary of Concepts 150
Exercises 153

6 Cultivating Creativity · 159
Some Problems with Puzzles 159
Extending Memory Beyond Miller's Magic Number 162
Unreasonable Effectiveness of Logic 165
Wiles Waiting to Prove Fermat's Theorem at Last 169
Beyond the Brain's Illusions 171
Creating Lives 172
The Apple of the Mind's Eye 180
Summary of Concepts 182
Exercises 183

Preview · 195

Epilogue · 201

Notes · 203

Acknowledgements · 207

Bibliography · 209

Index · 215

Preface

Note to Students

The ancient orator Horace (65-8 BC) once wrote, "Control your mind or it will control you." In today's society we are faced with more complex information and difficult decisions than ever before. Many people feel overwhelmed and powerless. One way to become less helpless, to gain control over your life, is to gain control over your own thinking. These modules are designed to impart skills for critically evaluating this barrage of arguments and opinions, data and statistics, fallacies and alternate realities. Instead of being overcome by the chaos, *Thinking Matters* aims to equip you with the presence of mind to comprehend it and to create order out of it.

Any course should do more than teach information. In nearly every field, 'facts' quickly become obsolete. The goals of this text are to help you:

- to solve problems more efficiently and creatively;
- to make logical deductions more fluidly and to expose fallacies more effectively;
- to identify moral principles used in ethical and legal debates and to critique them and their applications;
- to distinguish science from pseudo-science, to calculate probabilities, and to understand the logic of scientific testing;
- to code algorithms artfully and to comprehend the power and limits of computational thinking.

The text is punctuated with puzzles, experiments, and exercises to challenge and stimulate your curiosity. These may take the form of an inventory to be taken, a puzzle to be solved, or some questions to ponder. You are strongly urged to take the time to complete these tasks before reading on. These occasions can provide you with concrete, first-hand experiences of what is discussed more abstractly, but no less precisely, in the text that follows.

As a first example, suppose you are given the following premises:

> If Alice talks to the Cheshire Cat, she is mad. If Alice doesn't talk to the Cheshire Cat, she'll be lost. If Alice isn't mad, she will not be lost.[1]

Which of the following conclusions can be validly deduced about Alice?

(A) Alice is mad.
(B) Alice is not mad.
(C) If Alice doesn't talk to the Cheshire Cat, then she is mad.
(D) If Alice is mad, then she will be lost.
(E) If Alice isn't mad, then she talks to the Cheshire Cat.

Intuition is not always a reliable guide to judging validity. Therefore, mastering the techniques of logical deduction can help you *ace* such problems when you face them on standardized tests.

Have you chosen your answers?

In due course, you'll be introduced to logical terms and techniques. (For example, you'll quickly recognize that options (C) and (E) are logically equivalent by a law known as *contraposition*.)

I have to admit that I've purposely framed the question in a way to lull you into being a little logically lackadaisical. It turns out that there's more than one correct choice.

MARTIN GARDNER (1914-2010)

WARNING: Martin Gardner has turned dozens of innocent youngsters into math professors and thousands of math professors into innocent youngsters.
—PERSI DIACONIS

My lifelong fascination with logic, mathematics, magic, philosophy, and paradoxes began in fourth grade when I discovered Martin Gardner's first *Scientific American Book of Mathematical Puzzles and Diversions*. Reading the pages of its preface, I was captivated by a paragraph that shaped the way I approach the teaching of logic:

> [C]reative mathematicians are seldom ashamed of their interest in recreational topics. Topology has its origin in Euler's analysis of a puzzle about crossing bridges. Leibniz devoted considerable time to the study of peg-jumping... David Hilbert, the great German mathematician, proved one of the basic theorems in the field of dissection puzzles. The late Alan M. Turing, a pioneer in modern computer theory, discussed... the 15 puzzle in an article on solvable and unsolvable problems... [A] section of Einstein's bookshelf was stocked with mathematical games and puzzles. The interest of these great minds in mathematical play is not hard to understand, for the creative thought bestowed on such trivial topics is of a piece with the type of thinking which leads to mathematical and scientific discovery.[2]

For a quarter of a century, before the era of social media made creating communities of shared interests effortless, Gardner created a network of mathematicians (from John Conway to Stanisław Ulam), artists (from M. C. Escher to Salvador Dali), computer scientists (from Donald Knuth to Douglas Hofstader), magicians and skeptics (from Persi Diaconis to James Randi), novelists (from Isaac Asimov to Vladimir Nabokov), physicists (from Albert Einstein to Roger Penrose), philosophers (from Rudolf Carnap to Robert Nozick), logicians (from Lewis Carroll to Raymond Smullyan), and the list now includes you, if you so choose.

In retrospect, I have come to realize that Gardner's magical preface has become a map of my path as a logician. While pursuing a Ph.D. at UCLA I often introduced Gardner's recreational topics into the teaching of logic. As a result of this success, I was invited by Donald Kalish to co-author a second edition of the classic textbook by Kalish and Montague *Logic: Techniques of Formal Reasoning (Second Edition)*. Unwilling to choose between mathematics and philosophy, I took courses in logic in the Math Department from Herbert Enderton, C. C. Chang, and Donald (Tony) Martin, while taking courses in logic in the graduate program in Philosophy. My Ph.D. on the logical paradoxes was the last dissertation directed by Alonzo Church (whose fourth Ph.D. student was Alan Turing). Perhaps the most rewarding of all was to have my collaborative research discovering fractal images in the semantics paradox discussed by Ian Stewart in a descendant of Gardner's legendary column.

The idea of this book is not to intimidate you. Rather, it is to instill in you a healthy confidence in your ongoing process of critical thinking. It is not meant to confuse you with lots of jargon, nor to make you memorize tricks for passing standardized tests. Instead this book is designed to equip you with thinking tools which you can use for almost any intellectual endeavor for the rest of your life.

So try to relax and enjoy it. Remember that in a book that is trying to teach you to think, the effort of trying to think logically and critically counts for far more than getting the right answer. In a lot of cases, you may find that the 'right answer' is not right after all–or that perhaps you will find a better one than the one we found! So don't be afraid to use your head. If you can figure out how you are using it and see how to use it better, then this book will have accomplished its goal.

Our universe of thought emerges out of problems encountered in our everyday life just as our dreams emerge out of the events of the day. If we have a sense of wonder, the world may serve as a clue—a point of departure—suggesting questions that challenge us to consider ways of solving problems of a quite abstract and theoretical nature. Today our intellectual universe is so vast that in order to have some idea of its nature and extent, it is useful to divide it into smaller constellations. This procedure is useful provided that we keep in mind that these constellations are not separated by clear-cut barriers. This book is only a small map of many excursions that are possible, and it is based on the collected excursions of many pioneers who have explored before us. It is our hope that through reading and thinking your way through this book you will come to appreciate your thinking matters.

Note to Teachers

Critical thinking is hard to teach. It can be extremely challenging but it can also be extremely rewarding. One secret to teaching a successful course, I believe, is to use both your strengths and weaknesses. It is a good idea to begin and end the course with topics that are familiar to you. This allows you to concentrate on creating a community of learners at the beginning of the class and to inspire students to continue in the habit of critical thinking once the course is over.

To turn your weaknesses into strengths, I believe, it is always a good idea to be experimenting with new topics. This keeps the course, and you, engaged and evolving. If you're excited about learning or developing something new, your students are more apt to become enthusiastic learners and developers as well. One of my most memorable experiences in teaching a critical thinking course was when students organized a "class reunion" after the course was over because they didn't want the course, and the friends they had made, to end.

This *Thinking Matters* series is designed in modules so that you can easily structure your course according to your interests and those of your students. The modules are relatively independent and yet have interrelated themes. So think of the geometry of *Thinking Matters* not as a flat pentagon but as a fractal—a series of pentagonal stars on descending scales.

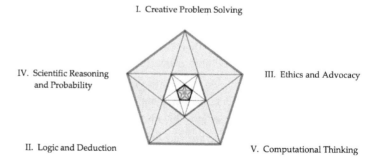

You can enhance your critical thinking abilities by adding to your repertoire of problem solving tools. The first module uses recreational puzzles and games to demonstrate handy tools that should be in such a toolbox–truth tables, inference rules and fallacies, logical deductions with directed graphs, Venn diagrams for syllogisms, probability trees, decision matrices, and algorithms. These methods, and schemas, are systematically developed in subsequent modules.

This module, *Critical Thinking as Creative Problem Solving*, can be read and understood by the students without additional lectures. Rather than lecturing, I like to begin the course with problem solving activities–e.g., students walking through a maze of bridges so they can discover Euler's theorem for themselves, a group problem solving exercise called "Lost on the Moon" demonstrating the benefits of "brainstorming" that can be turned into a group exercise turning "big data" into a technical report, or showing students how to ace some standard "intelligence tests" by learning a few heuristics. This first module is designed to put students in touch with, and to have confidence in, their own problem solving abilities.

The second module, *Critical Thinking as Logic and Deduction*, aims at imparting a toolbox of logical techniques. The Game of Deduction for the logic of propositions is introduced through an analogy with the Game of Chess. The series of events in *Through the Looking Glass* corresponds to moves in a chess game. Beginning at a Queen's pawn, Alice advances her way across the chessboard that is Wonderland, with seemingly nonsensical encounters with such colorful characters as Tweedledee and Tweedledum, Humpty Dumpty, and the White Knight. Their conversations and logical puns become playful, parodies of sense when viewed Through the Looking-Glass of symbolic logic. Upon reaching the 8th rank, Alice is promoted to a Queen and assists in checkmating the Red King in the middle of the chessboard. We supplement this system of logical deduction with the logic of quantifiers and relations which enable us to investigate smaller constellations of logical ideas–Aristotle's syllogisms, Venn Diagrams, Lewis Carroll diagrams, the logic of relations, and even some brief forays into modal logic and the logic of time.

The module *Critical Thinking as Ethics and Advocacy* prepares students to engage in debates and discussions in the public forum. The American philosopher John Dewey (1859 – 1952) warned, "Democracy has to be born anew every generation, and education is its midwife." Rather than merely learning a list of Latin names for fallacies or how to construct a Venn diagram, students acquire practical skills that enable them to advocate effectively for what they believe. Knowing

The function of education is to teach one to think intensively and to think critically. Intelligence plus character - that is the goal of true education.
    –MARTIN LUTHER KING, JR.
         (1929-1968)

these skills can free students to listen, more carefully and charitably, to those with whom they may disagree. In addition to learning about a dozen practical argument forms and their associated fallacies and various rhetorical skills and strategies, students will learn about the strengths and weaknesses of the most common ethical theories that are the grounds of arguing from principle. The fragility of democracy has never been more apparent. In the *Republic*, Plato argued that lawless democracies can descend into despotism. Winston Churchill might agree but added: "It has been said that Democracy is the worst form of government except all those other forms that have been tried from time to time."

The module *Critical Thinking as Scientific Reasoning and Probability* begins with the fallacies of theory testing committed in pseudo-scientific reasoning. Especially today when unsound scientific reasoning is used to bolster all kinds of political agendas—from attacks on the science of global warming to promoting fear about vaccinations or conflating cultural wars with belief, or non-belief, in evolution—it is important to develop some sound principles to distinguish scientific from pseudo-scientific reasoning. The fallacies of theory testing are then reverse-engineered and turned into conditions for a good scientific test. Learning the basic laws of probability turns these conditions into the requirements for the confirmation of an hypothesis through logic of Bayesian reasoning. This module also contains an introduction to the elements of decision theory.

Finally, *Critical Thinking as Computational Thinking* is relatively new for a course on critical thinking. Computational thinking is a set of problem-solving skills that arose from thinking about the nature of computation. Included in this skill set are constructing and "debugging" algorithms, organizing algorithms into simulations or computer models, and reverse-engineering algorithms to figure how they work or to discover their weaknesses or limitations. Computational thinking also encompasses design issues and the interface between computers and humans. This module brings us full circle to the first module on problem solving. Computational thinking will be as important in the 21st century as reading, writing and arithmetic were in earlier centuries.

Today, relevance to contemporary life is often demanded in logical studies. We have nevertheless found that historical and theoretical aspects of logic provide an intriguing and indispensable background. We have chosen not to shun theoretical topics in our desire to be practical. The examples in the text have been chosen from the headlines over the long period of time this material has developed and matured. Some attempts have been made to change dated examples. However, classic examples, such as blunders in televised presidential debates, remain. These blunders can be viewed by each new generation, perhaps informing the continuing dialogue about how democracy can, and should, be debated in the media in ways that build, rather than splinter and isolate, diverse groups into a more informed, participatory, and democratic citizenry. Other examples, such as the impeachment of President Richard Nixon, the vital importance of immigration in the face of the growing threat of World War II for America's excellence of higher education, the OPEC oil crisis, which has today become the looming threat of global warming—all have an unanticipated contemporary relevance. Paraphrasing the philosopher George Santayana (1863-1952), "those who cannot think critically about past mistakes are condemned to repeat them."

The text has several pedagogical features. Each chapter begins with a story or puzzle to introduce the subject that follows. Each section develops a cluster of basic concepts and sets forth various formal or informal techniques. The chapters end with a summary of the new concepts, and exercises are provided at the end of each section. The exercises are arranged into three groups. The first group consists of straight-forward exercises designed to help students discover whether they have understood the basic concepts of the section. The second group of exercises further develops and reinforces the concepts introduced in the chapter, and the third group consists of open-ended exercises designed to stimulate further interest in the subject. Inevitably as a teacher, what you ultimately impart to your students—and what they will remember most about the course—is what you model in class. Many of the values of critical thinking are the sort that must be 'caught' as you try your best to practice what you teach. Good luck!

A photograph is a most important document, and there is nothing more damning to go down to posterity than a silly, foolish smile caught and fixed forever.
—MARK TWAIN (1835-1910)

Progress, far from consisting in change, depends on retentivenesss. Those who cannot remember the past are condemned to repeat it.
—GEORGE SANTAYANA (1863-1952)

1

The Thinking Reed

As one grows older, one sees the impossibility of imposing your will on the chaos with brute force. But, if you are patient, there may come that moment when, while eating an apple, the solution presents itself and politely says, 'Here I am.' —ALBERT EINSTEIN (1879-1955)

Aristotle said that philosophy begins in wonder. Bertrand Russell, one of the most famous philosophers of the 20th century, also related philosophy and wonder:

> *Philosophy, if it cannot answer so many questions as we could wish, has at least the power of asking questions which increase the interest of the world, and show the strangeness and wonder lying just below the surface even in the commonest things of daily life.*[1]

It is appropriate then to begin this philosophy book on logical and critical thinking by puzzling about our ability to think.

The Wonder of Thinking

Wonder and puzzlement can arise in the context of everyday experience. To illustrate this point, here's a story about a five-year old named David, recounted in Gareth Matthews's *Philosophy and the Young Child*:

> David worries about whether an apple is alive. Is the apple on the table alive? David is puzzled. If it is alive, then when we eat it, we eat something that is alive. If it isn't, how does it differ from an apple still hanging from a tree?[2]

David's solution appears to be a creative compromise. He says that the apple is alive when it's on the ground but not when it has been brought into the house! We laugh at David's naive solution, since we, of course, all know the answer to David's puzzle–or do we?

One common way to attempt to answer the question of whether something is alive or not is by listing several life functions, such as digestion, reproduction, locomotion, etc. An organism is alive if it is

When an apple ripens and falls, why does it fall? Because of its attraction to the earth, because its stem withers, because it is dried by the sun, because it grows heavier, because the wind shakes it…?
  War and Peace
—LEO TOLSTOY (1828-1910)

capable of performing a sufficient number of these functions. But it's clear that David doesn't have this functional approach in mind.

We can get insight into David's approach if we consider some cases. Consider flowers. When we cut flowers in the garden and bring them inside, we put them in a vase with water to keep them alive. But we don't say we are keeping apples alive by putting them in the refrigerator. Instead we say we're keeping them fresh. Perhaps we can think of an apple's being alive not in terms of satisfying a number of life functions but rather in terms of its natural life cycle. An apple contains seeds and nourishment for seeds. If an apple is left on the ground, a seed may germinate and produce a little apple tree. This little tree may in turn produce apples of its own. The cycle continues.

Perhaps death occurs when a life cycle is interrupted, for example, when a flower is uprooted from its soil, when a caterpillar is prevented from spinning its chrysalis, or when an apple is brought indoors so that its seeds can no longer germinate in the soil. Perhaps David's solution is more than just a creative compromise: it is a child's fresh and ingenious solution to a very old and persistent philosophical puzzle.

In addition to showing how puzzlement can arise in everyday life, this example contrasts with the way many 'educated' adults have come to relate to their own thinking.

Almost without exception, children love puzzles—unfortunately, they are sometimes taught to relegate wonder to a position of unimportance. Wondering is a child's way of encountering his or her world. Children naturally love to *solve* puzzles—it is a form of play, a way of interacting with the world around them. When I used to invite my young daughter to think, she would shower me with an abundance of questions and answers. Sometimes when I ask adults to wonder about philosophical problems, they will give me a shower of excuses. They have come to regard thinking as something that philosophers, scientists, or super sleuths mysteriously do for them. How did this sad situation come about? I suggest two reasons.

First, criticism, prejudice, ridicule, and insecurity have stifled much of our willingness to wonder. People often find it easier to ridicule the imperfect thinking of others than to try to think for themselves. This tends to discourage the habit of thinking itself. After all, if I can't produce a masterpiece, why try?

Secondly, as adults we think we have enough problems. We try to avoid them or, failing that, we try to dispatch with them as soon as possible. We have lost the joy in solving problems because they confront us with ever increasing frequency–your car breaks down, you've had a fight with someone you care about, you're at a restaurant and you've forgotten to bring the coupons you cut out and the cash machine won't return your bank card as you're trying to catch the train. Yet this common attitude of avoidance can hardly lead to improving one's problem solving skills. Indeed, it may even lead to worse problems. As James Adams observed in *Conceptual Blockbusting*:

> Few people like problems. Hence the natural tendency in problem-solving is to pick the first solution that comes to mind and run with it. The disadvantage of this approach is that you may run either off a cliff or into a worse problem than you started with.[3]

So even if we try to avoid problems as best we can, problems, undoubtedly, will find us. There is no good alternative to finding ways of improving our problem solving abilities.

Just as wondering is a joyous experience for children, I believe that it can, and should, be a similar positive experience for us as we grow older. One goal of this course is to help you to rediscover the joy of problem solving. It is my conviction that problem solving is a skill that can be taught and improved with practice. Knowledge of the principles and strategies of problem solving can make us better at it. We may even come to enjoy solving problems for fun rather than out of necessity.

Thought & Creativity

When you think of the word 'critical' what others come to mind? Here are some synonyms given by *Webster's*: *captious, carping, cutting, disparaging, hair-splitting, nit-picking, caustic, corrosive, sarcastic*. To be critical has come to have quite a negative connotation. Yet if we go back to its Greek roots, the word 'critical' comes from the Greek word *kritikos* meaning 'able to judge or discern.' To be critical originally had the positive connotation of being equipped to form an educated opinion.

Before launching into our study of problem solving, it is perhaps useful to deflate two common misconceptions about the relationship between creativity and critical thinking.

The first misconception is that critical thinking is antithetical to creative thinking, an artistic and intuitive process. It has become increasingly difficult these days to avoid knowing that the brain consists of two nearly identical (anatomically speaking) hemispheres with different functions. But we can raise two critical questions. Is it so? So what if it is?

Consider the second question first. Many in the Human Potential Movement have blamed the ills of our society on our entire culture being 'too left-brained.' Society has just become too rational and verbal and scientific. They insist that we should suspend rational thinking and seek to rediscover reality by relying solely on intuition. This offers a simple solution to a complex problem. If the ills of society are caused by being 'too left-brained,' all we have do is to start being more 'right-brained.' Another appeal is this: it suggests that our scientific, logical culture has diminished what it is to be fully human by ignoring an entire half of the brain. If we start to become more 'right-brained,' we will use our whole brain and become more whole persons.

According to popular accounts, the two hemispheres have the following properties:

LEFT BRAIN	*RIGHT BRAIN*
verbal	visual
scientific	artistic
convergent	DIVERGENT
rational	*Intuitive*

Left-Brain and Right-Brain Dichotomies

Two points of logic need some attention right away. First, merely contrasting notions like rational and intuitive tends to stereotype the meaning of each and to discourage remembering cases of thinking in which both qualities are present. Take scientific reasoning, for example. Such thinking involves not only drawing logical implications of a hypothesis but also the creative construction of hypotheses and theories. Albert Einstein claimed to use both visualization and kinesthetic cues regularly in his work as a scientist:

> The words or the languages, as they are written or spoken, do not seem to play any role in my mechanism of thought. The psychical entities which seem to serve as elements in thought are certain signs and more or less clear images which can be 'voluntarily' reproduced and combined.... The above mentioned elements are, in my case, of visual and some of muscular type. Conventional words or other signs have to be sought for laboriously only in a secondary stage, when the mentioned associative play is sufficiently established and can be reproduced at will.[4]

Secondly, it should be clear that merely *locating* the various brain functions in a particular hemisphere of the brain does little toward improving our creativity. Many have seized upon the right-left brain research as a metaphor rich in meaning. But how much does knowing *where* invention is supposed to occur in the brain tell us about *how* to go about improving our ability to invent?

Let's return to our first question, 'Is it so'? That is, do the scientific studies really show that the right-brain is visual, intuitive, and artistic as *opposed* to the left-brain which is verbal, analytic, and scientific? The early split-brain research of Nobel Prize winner Roger W. Sperry has undergone a number of critical reappraisals.[5] We now know, for example, that a wholesale assignment of language functions to the left hemisphere is oversimplified. Processing of vowel sounds and access to the meanings of words, for example, now appear to be operations performed in both hemispheres.[6]

Besides qualifications of this sort, the correlations between the kind of thinking supposedly involved and the hemisphere employed are very low. The original experiments on split-brain research were performed on patients already disabled with general physical disorders, such as epilepsy. Such patients should not automatically be assumed to accurately represent the brain processes of non-disabled patients. Furthermore, poor performance of a task resulting from a localized brain injury does not imply that the injured portion of the brain performed the task. If a radio stops broadcasting when the knob is knocked off, that doesn't prove that the knob did the work of broadcasting. So, despite what many people might like to believe, the sharp division of thought processes into 'left' and 'right' brain activities is not unequivocally supported by scientific evidence.

Why then do such polarities and dichotomies appeal to us in the first place? Perhaps one reason is the relative ease with which we can categorize the world into such polarities. If it's easy to classify people into two types: rational versus intuitive, type A or type B, sensitive or scientific, then it's tempting to suppose that these polarities are fundamental dimensions of the human personality.

But before accepting that so quickly, we should try an exercise suggested by E. H. Gombrich in *Art and Illusion*.[7] Consider the following polarity–Ping versus Pong. Eccentric perhaps, but nonetheless it's remarkably easy for us to sort the world according to this polarity. What is a fork? Definitely, Ping. What about a spoon? Pong, undoubtedly. What about the ocean? Pong, of course, and a star would be Ping. Isn't it easy to categorize things according to Ping versus Pong, Yin versus Yang, Male versus Female, Left Brain versus Right Brain? The intended moral here is perhaps obvious: ease of classifying proves nothing. Most any polarity, even the silly Ping versus Pong, provides an intuitive way of mapping the world.

Perhaps this is enough to dispel the first misconception about the relationship between critical and creative thinking. Critical and logical thinking is by no means divorced and unconnected from creative and intuitive thinking. The two kinds of thinking are, in fact, married in almost every kind of intellectual endeavor.

Creative Thinking & Logical Inference

A second misconception about the relationship between critical and creative thinking is this: creative thinking is merely a matter of logical thinking–better mental programming. Alan Turing, the grandfather of Artificial Intelligence, confidently predicted in 1950 that "at the end of the century the use of words and general educated opinion will have altered so much that one will be able to speak of machines thinking without expecting to be contradicted."[8]

What about research into artificial intelligence (AI)? Is there any significant difference between human thinking and computer processing? To be more specific, let's consider a puzzle known as a *transitive inference problem* and the solution to such puzzles posed by AI researchers. You are a sales person and ask a customer what brand of stroller he would like to buy. The customer, a new father driven to distraction by his newborn, replies:

1. ***Apricots*** IS BETTER THAN ***Baby Time***.
2. ***Cuddles*** IS NOT AS GOOD AS ***Diaper Day***.
3. ***Fussy Face*** IS WORSE THAN ***Grouch Pouch***.
4. ***Cuddles*** IS BETTER THAN ***Grouch Pouch***.
5. ***Fussy Face*** IS BETTER THAN ***Apricots***.

Given these preferences, can you quickly determine which brand the harried dad likes the best? Due to the limits of short term memory such problems are sometimes difficult to solve without some notational aid. Before reading on, take some time to solve the problem on your own. What follows will be more meaningful and interesting if you take a moment to have some first-hand experience.

Now compare your method of solving this problem with the following 'memory efficient' problem-solving program proposed by some AI researchers:

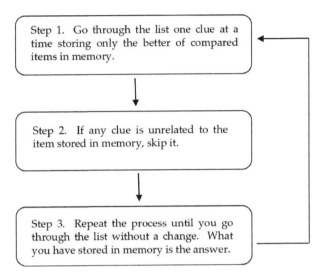

Mental Software for Transitive Inference Problems

Let's apply this program to the above puzzle. Looking at the first clue:

(1) ***Apricots*** IS BETTER THAN ***Baby Time***.

We place ***Apricots*** in memory. Going on to line (2), we notice that ***Apricots*** is not mentioned, so it will be skipped for now. The same applies to lines (3) and (4). Line (5), however, states:

(5) ***Fussy Face*** IS BETTER THAN ***Apricots***

and so we replace ***Apricots*** with ***Fussy Face*** in our memory.

Returning to the beginning, we skip lines (1) and (2) because they do not mention ***Fussy Face*** Moving on the line (3) we have:

(3) ***Fussy Face*** IS WORSE THAN ***Grouch Pouch***.

so we replace ***Fussy Face*** with ***Grouch Pouch*** in our memory. Now line (4) states:

(4) ***Cuddles*** IS BETTER THAN ***Grouch Pouch***.

and so ***Grouch Pouch*** is out and ***Cuddles*** is in our memory. Line (5) is not relevant so we return to line (1) which is also not relevant to the item in memory. Line (2), however, states:

(2) ***Cuddles*** IS NOT AS GOOD AS ***Diaper Day***.

So we replace **Cuddles** in memory with **Diaper Day**. Going through the list one more time, we see that nothing changes the item stored in memory. It also happens that all the clues have been checked. Distracted dad likes **Diaper Day**.

Now let's see if your problem solving abilities have improved with the aid of this mental program. Here's another transitive inference problem. This time you're selling cars and you need to determine the color the customer most prefers given the following preferences:

Aquamarine IS NOT AS GOOD AS *Tangerine*.
Burgundy IS BETTER THAN *Canary Yellow*.
Cyan IS NOT BETTER THAN *Aquamarine*.
Magenta IS BETTER THAN *Burgundy*.
Tangerine IS WORSE THAN *Canary Yellow*.

Did you solve the problem more efficiently this time? Was it due to superior mental programming? Or perhaps you were more familiar with transitive inference problems? Did it help to have a strategy that was less susceptible to errors of short term memory?

Thinking critically about this experiment, what has it really shown? Have we demonstrated that thinking is nothing but computation–as some more radical AI researchers wish to conclude? Has the computer program simulated human problem solving? Or perhaps is it that you have become more like the computer? (The answer, by the way, is Magenta.)

But wait a minute. Perhaps you, unlike a computer, have already begun to question the neat little mental algorithm you have just received. How often do you encounter a real-life problem such as this? Perhaps you began to question whether human preferences are linearly ordered–an assumption required for a solution to the puzzle. Can you construct a transitive inference problem in which this mental program could lead you down two quite different paths? How do you know you've reached the global maximum and not merely a local maximum? Now you're wondering about the limitations of the neat AI program designed to improve your intelligence and whether it can be significantly improved. Perhaps you're wondering if the program simulates how you originally solved this problem. Could it be (you wonder) that this wondering sets us human beings apart from computers?

Let's sum up the discussion. We need to be critical about popular (mis-)conceptions about the nature of logical and critical thinking. Psychological research, at any rate, does not support the Human Potential rhetoric that attempts to relegate creative thinking to a mystical realm beyond logical thinking, nor has the recent research in artificial intelligence reduced creative thinking to mere efficient mental programming. Our ability to think and to wonder about our own thinking, it seems, remains an intriguing puzzle of its own.

Throughout our history, we human beings have considered ourselves unique among the creatures of the earth. Sometimes we have gloried in our reason as Shakespeare's Hamlet proclaimed, *"What a piece of work is man, how noble in reason"*, and sometimes we have humbly acknowledged we have been formed from the dust of the earth, as Charles Darwin concluded in *The Descent of Man*, *". . . man with all his noble qualities. . . still bears in his bodily frame the indelible stamp of his lowly origins."* And sometimes we have acknowledged the paradox of our greatness and lowliness, as Pascal observed in his *Pensées*: *"Man is but a reed, the most feeble thing in nature, but he is a thinking reed."*[9] But noble or lowly, or both, we have always thought ourselves unique because of the wonders of the human mind.

Summary of Concepts

Philosophy begins in wonder. Our ability to think critically and creatively is an intriguing wonder of its own. CRITICAL THINKING doesn't refer to mere negative thinking, but rather to the positive acquisition of the THINKING SKILLS that can improve your overall thinking abilities, skills that can be improved with thoughtful effort.

Psychological research about "left and right brain thinking" does not support the mystical view according to which critical and logical thinking is simply opposed to creative and intuitive thinking. CREATIVE THINKING involves critical thinking and critical thinking can be creative as well.

Neither does recent research in ARTIFICIAL INTELLIGENCE and cognitive science support the reductionist view that creative thinking is nothing but better MENTAL PROGRAMMING. Thinking that discovers the limits of a mental program can itself be a creative endeavor not reducible to just more mental programming.

What a piece of work is a man! How noble in reason, how infinite in faculty! In form and moving how express and admirable! In action how like an angel, in apprehension how like a god! The beauty of the world. The paragon of animals... —SHAKESPEARE (1564-1616)
Hamlet, Act II, Scene 2
[1599 – 1601]

We must, however, acknowledge, as it seems to me, that man with all his noble qualities, with sympathy which feels for the most debased, with benevolence which extends not only to other men but to the humblest living creature, with his god-like intellect which has penetrated into the movements and constitution of the solar system — with all these exalted powers — Man still bears in his bodily frame the indelible stamp of his lowly origin.
—CHARLES DARWIN (1809-1882)
The Descent of Man,
Chapter XXI [1871]

Man is a reed, the weakest of nature, but he is a thinking reed. It is not necessary that the entire universe arm itself to crush: a vapor, a drop of water suffices to kill him. But if the universe were to crush him, man would still be nobler than what kills him, because he knows that he dies and the advantage that the universe has over him, the universe does knows nothing.
—BLAISE PASCAL (1623-1662)
Pensées, 347 [1670]

Exercises

Group I

1. **Critical Thinking**

   Write down a list of words that come to mind when you hear the word 'critical.' Now, without thinking too hard, make a 'top ten' list of critical thinkers. What does the word 'critical' mean to you? Now look up the Greek roots of the terms and find out what it meant classically. How does the classical meaning of the word 'critical' different from its contemporary connotations?

	Word Associations	Top Ten List of Critical Thinkers
1.		
2.		
3.		
4.		
5.		
6.		
7.		
8.		
9.		
10.		

2. **Thinking Critically about Critical Thinkers**

   Go back and review your list of critical thinkers. Does it reveal any stereotypes you might have about critical and logical thinking? Are there any women, racial or ethnic minorities, or disabled people on your list? What about athletes or architects, social reformers or saints, children, poets or painters, or musicians or mystics? Now go back to your top ten list and intentionally try to include a diversity of critical thinkers. Perhaps our imagination of what it means to be a critical thinker is itself in need of some critical reflection.

3. *Creating Inventories for Change*

People everywhere have asked what Rodin's *Thinker* is pondering? The answer many children have come up with is a good one: he's thinking "Where did I leave my clothes?" The stark nakedness of the thinker is revealing: Rodin's thinker is not a flabby sedentary couch potato–he's a person of action. Rodin's own reflection on his masterpiece:[10]

The Thinker
—Auguste Rodin (1840-1917)

Guided by my first inspiration I conceived another thinker, a naked man, seated on a rock, his fist against his teeth, he dreams. The fertile thought slowly elaborates itself within his brain. He is no longer a dreamer, he is a creator.

Take an inventory of your own creative and critical thinking skills. What do you do well and what would you like to change?

Things I'd Like to Keep	Things I'd Like to Change
1. _____	_____
2. _____	_____
3. _____	_____
4. _____	_____
5. _____	_____
6. _____	_____
7. _____	_____

Now go back and look at your expanded list of critical thinkers and critical thinking skills. Choose a mentor to teach you how to develop your skills. How did they fire up their passion, how did they practice their skills, in what venues where these skills performed? Now think of ways in which you could structure into your life some mini-clinics to start incorporating the development of these skills into your daily life. Now just do it!

Group II

4. Hill-Climbing

Sherlock Holmes is trying to find the top of a hill in a thick fog. He can't see where the top is. But at least he can see which direction is uphill. So he does the logical thing. He walks in that direction. If he walks uphill long enough, he's sure to get to the top. 'Hill-climbing' is a progressive search that proceeds by a series of improvements. Step by step, the progressive search homes in on its objective. Hill climbing is a good metaphor for many human activities. Short, forward steps can lead to creative results.

Think of three examples in which hill-climbing will lead to a satisfactory solution of a problem. Then give three examples in which this method most likely fails to produce the satisfactory results.

5. Create Your Own Transitive Inference Problem

(A) Go back to the transitive inference problem and create two of your own problems. Make sure that the problem is 'attractive'—not too easy, but not too difficult either.

(B) Compare the two problems, which is 'better'? Test the AI program on your problems to see if it works.

(C) Now create a hybrid of the two problems in which the method of 'hill-climbing' method gives you different answers when you begin at the top of your list and when you begin at the bottom.

6. Transitive Dog Trainers

There are four trainers and five types of dogs at the dog rescue. Algernon trains Chinese Shar-Peis and Poodles. Gisselle trains German Shepherds and Chinese Shar-Peis. Jasmine trains German Shepherds and Bulldogs, and Laurette trains Rottweilers and Bulldogs. It happens that Bulldogs are easier to train than Chinese Shar-Peis, and Poodles are easier to train than German Shepherds. German Shepherds are easier to train than Bulldogs, and Rottweilers are harder to train than Chinese Shar-Peis. Which trainer trains the most difficult dogs? If Chinese Shar-Peis are far more difficult to train than Bulldogs, and German Shepherds and Poodles are about as easy to train as Bulldogs, which trainer trains the least difficult dogs

7. *A Metaphorical Optical Illusion*

Now compare your solution in 5 with attempts to extend the perceptual interpretation from a corner of the *Penrose Triangle* to the entire triangle. Use this metaphor to improve your creative solution to problem.

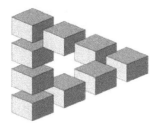 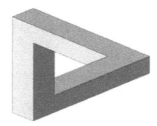

Penrose Triangles

Group III

8. *Worthless Propositions*

Here's an excerpt from Carl Sagan's famous essay, "The Fine Art of Baloney Detection":

> What skeptical thinking boils down to is the means to construct, and to understand, a reasoned argument and—especially important—to recognize a fallacious or fraudulent argument. The question is not whether we like the conclusion that emerges out of a train of reasoning, but whether the conclusion follows from the premise or starting point and whether that premise is true.[11]

Among Sagan's few simple rules:

- Propositions that are not testable are worthless–you have to be able to check the assertions out.
- There must be substantive debate–it isn't enough simply to attack your opponent's character.
- If there is a chain of argument, *every* link in the chain must work, including the premises–not just most of them.
- Arguments from authority carry little weight–'authorities' have made mistakes in the past and will do so again.

Let's think critically about Sagan's rules.

(A) Find examples in which each of Sagan's rules is helpful for distinguishing between 'baloney' and 'real science'.
(B) Try to find examples in which people find it useful to violate Sagan's rules. For example, under what circumstances do

political advisors advise their candidates to "go negative" and attack the opponent's character? Should we accept Sagan's advice, as an authority on scientific reasoning, not to use arguments from authority?

(C) Now consider Sagan's rule that "propositions that are not testable are worthless". This is a popular statement of *scientism*–the philosophical view that only scientific statements are meaningful. Is this proposition self-referentially consistent?

9. *Transcending Transitive Inferences*

The ordering of our preferences can exhibit various properties of relations.

- A preference ordering on a set S is *irreflexive* if for every x in S x is not preferred to x.
- A preference ordering on a set S is *transitive* if for every x in S whenever x is preferred to y and y is preferred to z, x is preferred to z.
- A preference ordering is *strict* if it is both irreflexive and transitive.
- A preference ordering is *connected* in a set S if for any two distinct members x and y of S, either x is preferred to y or y is preferred to x.
- A preference ordering is *linear* if it is both strict and connected.

(A) Show that if the customer's preference ordering fails to be transitive, the AI program for solving transitive inference problems can be thrown into an infinite loop and never halt.

(B) Show that even if the customer's preference ordering is transitive, it may not be linear, in which case the AI program may not give the optimal or best solution.

(C) Show that if the customer's preference ordering is not connected, the AI program could give you a different answer if you started at the bottom of the list of clues instead of at the top.

10. *Multiply Connected Transitive Inferences*

A consumer interest group rates the fruits and vegetables in a supermarket's produce section. The rating scale is A, B, C, D, E, where A is the highest rating and E is the lowest rating. An E rating indicates spoilage. Two letter ratings are consecutive if and only if they are adjacent in the alphabet. The fruits and vegetables that the store carries are as follows: eggplants, grapes, kiwis, tomatoes, plums, and raspberries.

- The store's eggplants are rated higher than its kiwis.
- The store's grapes and plums are given ratings that are consecutive.
- The store's tomatoes and raspberries are given ratings that are consecutive.
- The store is given a higher rating for its grapes than its plums.

(A) If the store is given the same rating for its eggplants as its tomatoes, and if the raspberries are given a rating that indicates spoilage, which one of the following must be true?

   a. The store's grapes are given a B.
   b. The store's kiwis are given a D.
   c. The store's kiwis are given an E.
   d. The store's plums are given a B.
   e. The store's plums are given a C.

(B) If none of the fruits and vegetables are given failing ratings, and the grapes are given a higher rating than either the tomatoes or the raspberries, which one of the following must be true?

   a. The store is given exactly one A.
   b. The store is given exactly one B.
   c. The store is given exactly two Bs.
   d. The store is given at least one B and at least one C.
   e. The store is given at least one C and at least one D.

(C) If the store is given a higher rating for its plums than its eggplants and is given a higher rating for its eggplants than either its tomatoes or its raspberries, which one of the following would allow all six of the ratings to be determined?

   a. The store's plums are rated D.
   b. The store's tomatoes are rated D.
   c. The store's kiwis and tomatoes are given the same rating.
   d. The store's kiwis and raspberries are given the same rating.
   e. The store's kiwis are rated higher than its raspberries.

(D) Assume that the store is given a higher rating for its plums than its eggplants. Also assume that the store is given a higher rating for its kiwis than its tomatoes. Exactly how many of the store's ratings can be determined?

   a. two
   b. three
   c. four
   d. five
   e. six

(E) If the store's tomatoes are rated lower than the plums and also consecutive to the rating of the plums, and if the ratings of the plums and raspberries are not identical, then which one of the following must be true?

    a. An A and a C rating are received by the store.
    b. An A and a B rating are received by the store.
    c. A D and a B rating are received by the store.
    d. A D and an E rating are received by the store.
    e. An E and a B rating are received by the store.

(F) Which of the following, if true, would guarantee that whenever the store's plums are rated higher than its eggplants, then there must be at least one spoiled item?

    a. The eggplants are rated higher than the tomatoes.
    b. The plums are rated higher than the tomatoes.
    c. The tomatoes are rated higher than the plums.
    d. The eggplants are rated higher than the raspberries.
    e. The kiwis are rated higher than the raspberries.

After you've worked with the problem think of a model to help you to solve it better. For example, you might think of a slide rule in which a set of disconnected rulers that can slide together in various ways.

11. *Minds, Machines, and Gödel's Disjunction*

(A) The following claims made by researchers in a strain of artificial intelligence (AI) known as 'Strong AI' are quoted in John Searle's *Minds, Brains, and Science*.[12]

- "Herbert Simon of Carnegie-Mellon says we already have computers that can think. ... Simon's colleague Alan Newell says we have now discovered ... that intelligence is just a matter of physical symbol manipulation."
- "Marvin Minsky of MIT says that the next generation of computers will be so intelligent that we will 'be lucky if they are willing to keep us around the house as household pets.'"
- "John McCarthy, [who with Minsky coined the term 'artificial intelligence,'] says that even "machines as simple as thermostats can be said to have beliefs.""

Torkel Franzén in *Gödel's Theorem: An Incomplete Guide to Its Use and Abuse* [2005] discusses the claims advanced by J. R. Lucas [1961], challenged by Douglas Hofstader in *Gödel, Escher, Bach: An Eternal Golden Braid* [1979], and updated by Roger Penrose in the *Emperor's New Mind* [1989] and *Shadows of the Mind* [1994].[15] [16]

(B) The essence of Lucas's Argument is the following claim:

> "Gödel's theorem states that in any consistent system which is strong enough to produce simple arithmetic there are formulas which cannot be proved in the system, but which we can see to be true."[17]

No matter how complicated a machine we construct, it will correspond to a formal system, which, in turn, will be subject to a Gödelian construction for finding a statement that is unprovable in that system, assuming the formal system is consistent. What paradoxes arise when a formal system asserts and proves its own consistency?

(C) Douglas Hofstader in *Gödel, Escher Bach: An Eternal Golden Braid* [1979] popularized Gödel's ideas and promotes his own pro-AI agenda by means of a metaphor: if self-reference can arise from a sufficient complex formal system, then self-consciousness should be able to arise out of a sufficient complicated neuronal formal system.

> "The other metaphorical analogue to Gödel's Theorem which I find provocative suggests that ultimately, we cannot understand our own minds/brains.... All the limitative theorems of mathematics and the theory of computation suggest that once your ability to represent your own structure has reached a certain critical point, that is the kiss of death: it guarantees that you can never represent yourself totally."[18]

Critically evaluate Hofstader's metaphor.

(D) Penrose updates the Lucas argument in an attempt to show that the aspirations of strong AI are doomed to failure. Penrose then conjectures that a non-computational extension of quantum mechanics will someday provide a theory of consciousness. What are some problems with such promissory speculations?

(E) Gödel's own remarks on the subject (in his unpublished 1951 Josiah Willard Gibbs Lecture at Brown University, see *Collected Works: Volume III: Unpublished Essays and Lectures*, edited by Feferman et al.) are characteristically more cautious:

> "The human mind is incapable of formulating (or mechanizing) all its mathematical intuitions. i.e.: If it has succeeded in formulating some of them, this very fact yields new intuitive knowledge, e.g. the consistency of this formalism. This fact may be called the 'incompletability' of mathematics. On the other hand, on the basis of what has

been proved so far, it remains possible that there may exist (and even be empirically discoverable) a theorem-proving machine which in fact is equivalent to mathematical intuition, but cannot be proved to be so, nor even proved to yield only correct theorems of finitary number theory.

The second result is the following disjunction: Either the human mind surpasses all machines (to be more precise: it can decide more number-theoretic questions than any machine) or else there exist number theoretic questions undecidable for the human mind."[19]

In what ways is Gödel's disjunction less subject to the criticisms that apply to the above authors?

2

"Eureka!" Problem Solving Heuristics

To solve a problem is to make a discovery: a great problem means a great discovery, but there is a grain of discovery in the solution of any problem. Your problem may be modest; but if it challenges your curiosities and brings into play your inventive faculties, and you solve it by your own means, you may experience the tension and enjoy the triumph of discovery.
—GEORGE PÓLYA (1887-1985)

The very phrase "I have a problem" signals that something's wrong, that we're in trouble. As a result, we often try to ignore the problems, put off dealing with them, or to get rid of them as quickly as possible. But this attitude of avoidance often keeps us from finding creative solutions. And such an attitude deprives us of opportunities for intellectual development. When we start to solve problems successfully, when we know how to approach problems in a systematic way, we can come to regard problem solving as a pleasant and productive activity. What is problem solving anyway? And why is it so important?

Characterizing Problems

Our first problem is to define what a problem is! How can we characterize a problem in a useful and abstract way? Allen Newell and Herbert Simon in their classic work, *Human Problem Solving*, define a problem as follows:

> A person is confronted with a problem when he [sic] wants something and does not know immediately what series of actions he can perform to get it [1].

In short, a problem exists when the given state (*what is*) differs from the goal state (*what is wanted*). A solution path to a problem is a series of operations that we can perform to change the given state into the goal state.

Suppose, for example, you're hungry. That's the given state. Your goal state is to alleviate your feelings of hunger. But perhaps you're also trying to lose weight; in that case, your goal state may be to live with the hunger until the next scheduled mealtime. *Defining the problem*—understanding what the given and goal state really are—is an important first step in problem solving.

If you're not on a diet, one simple solution to this problem might be to go to the refrigerator and have a snack. But what if you're not home? Then instead of raiding the refrigerator, you may have to go to a grocery store or a sushi bar. The solution to a problem may depend on a particular context in which the problem is to be solved. This context is called the *problem space*. The problem of finding a glass of water in your house is vastly different from that of finding a glass of water in the middle of the Sahara desert. The same problem in different problem spaces may require vastly different solutions.

Any problem also has *constraints*–limitations on what constitutes an acceptable operation as part of your solution path. Raiding your neighbor's pantry would probably not be an acceptable solution and so would be a constraint. Similarly, if you don't like eating raw fish (like me), you probably don't want to go to a sushi bar. Constraints limit the set of acceptable solutions. We can summarize these concepts in a diagram:

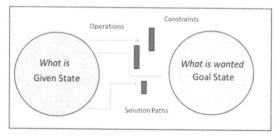

Let's apply these concepts to a classic problem studied by Duncker in 1945:

> Given a human being with an inoperable stomach tumor and rays which destroy organic tissue at sufficient intensity, by what procedure can one free him of the tumor by these rays and at the same time avoid destroying the healthy tissue which surrounds it? [2]

Let's see if we can use our terminology to characterize this problem.

- What is the given state? A patient with inoperable stomach tumor.

- What is the goal state? To remove stomach tumor from patient.

- What are the various operations that we can perform? Excise tumor with scalpel. Use rays that destroy organic tissue at sufficient intensity. Implant a plug to cut off the blood supply to the tumor.

- What are some constraints? Don't kill—or needlessly damage—the patient (you all know the saying "the operation was a success but the patient died.")

- What is the problem space? The modern medical environment.

PROBLEM: How can you use the rays to destroy the tumor without destroying the healthy tissue?

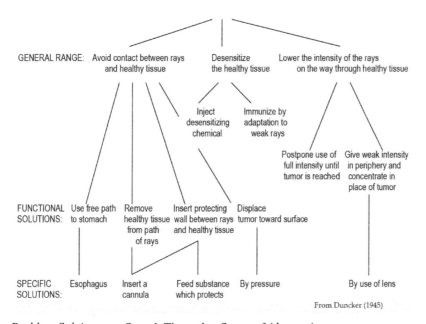

From Duncker (1945)

Problem Solving as a Search Through a Space of Alternatives

So according to Newell and Simon, problem solving is a search through a tree of alternatives. Improving problem solving therefore involves efficient pruning of the tree of alternatives. Better problem solving depends on such things as the problem solver's ability to:

- produce fruitful alternatives
- prune futile ones
- reject inferior options

- recognize and accumulate genuine improvements
- know when to backtrack and embark on a totally new approach
- know when to stop looking .

Our process of searching for a solution to the tumor problem using rays is diagrammed as a tree of branching paths through the problem space of alternatives–some of which lead to dead ends and others of which lead from the given state to the goal state.

Problem solving strategies, rules of thumb, which may help in this process are known as *heuristics*. (Later I'll explain the story behind this term which is derived from the Greek phrase "Eureka" meaning "I found it!") In the remainder of this chapter, we'll talk about seven heuristics that can be used to help you to become a more effective problem solver.

Isolate the Real Problem

One block to creative problem solving is the failure to isolate the real problem. Isolating the real problem involves correctly representing the given state and the goal state; adding constraints to sharpen the problem and locating the problem in the correct problem space.

Which figure doesn't belong to the set?

Considering this problem, many people construe the question to mean, "Find the one odd figure in a set of regular ones," and so answer figure 3 because it is circular. But if they then recheck their solution (as most people do), they are likely to notice that figure 1 is the only one with a square in the middle instead of a circle, that figure 2 is the only shaded out figure, that figure 4 is the only one with a double border. All of which may be frustrating. But this frustration usually leads to a closer examination of the problem. Exactly what is it that we are given and what is our goal?

We were asked, "Which figure doesn't belong to the set?" This could mean "find the *irregular* item among a set of *regular* ones" but it could also mean "find the *regular* item among the set of *irregular* ones". As noted above, every figure, except for figure 5, has something irregular about it. Figure 5 has a square border (which 3 out of the other 4 have); it has a circular interior (which 3 out of the other 4 have); its interior is also unshaded (as are 3 out of the remaining 4); and its border is not double (as are 3 out of the remaining 4). The answer therefore is figure 5.

Isolating the problem—clearly defining the given and goal states—is the first step toward improving your problem solving abilities.

Consider next a social problem, "How can we global warming crisis?" Solutions come quickly to mind–global adoption of alternative fuel technologies, use of mass transportation and driving electric cars, converting to solar power for heating and cooling, ending deforestation, cutting back on air travel, among other measures. In an historical case—the energy crisis that occurred in the early 1970s—some solutions to the crisis would have included our need to carpool, the need to end our dependence the oil cartels in the Middle East, and perhaps the need to develop solar-powered cars. During that OPEC oil crisis, it was proposed that cars with even numbered license plates could be fueled on even numbered days, and cars with odd numbered license plates could be fueled on odd numbered days. James Yorke, a mathematician who liked to think of himself as a philosopher, testified to the State of Maryland that the odd-even system of limiting gasoline sales would in fact only make the gas lines longer. Can you figure out why?

In any case, what's not so obvious here is: "What's the real problem?" Was the problem of the energy crisis a *technical* one of finding ways to conserve on gas, the *logistical* problem of distributing gasoline more fairly, the *political* one of wresting the power from the oil producing nations, or the *ecological* problem that the earth is running out of fossil fuels? It would be better to introduce more constraints, or to divide the problem into several separate problems.

The odd and even system of gas rationing led to longer lines at the pump as Yorke had predicted. From computer simulations, Yorke found that the odd-even system forced drivers to make more trips to the filling station to keep their tanks full all the time; thus the system increased the amount of gasoline sitting wastefully in the nation's automobiles at any moment rather than being distributed to those who needed it at the time.[3]

Another common mistake in problem solving is making unwarranted assumptions about the context or space in which the problem is to be solved.

The Nine Dot Puzzle is a classic illustration of this. Perhaps you saw it in elementary school. Can you, without taking your pencil off the paper, draw no more than four straight lines that pass through all nine dots as arranged below? (If you already know the standard solution, try to find a better one!

Nine Dot Puzzle: Cross 9 Dots with 4 Continuous Straight Lines

On their first attempts, many people unsuccessfully attempt to solve the puzzle like this:

These unsuccessful attempts illustrate the fact that we often tend to make unwarranted assumptions about the problem space in which the problem is supposed to be solved. Many people unconsciously limit themselves to solving the problem in the square outlined by the nine dots.

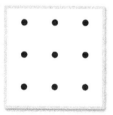

Imposing an Imaginary Boundary Unduly Limits the Problem Space

All the solutions that I know of—except one—involve expanding the problem space. In *Conceptual Blockbusting*, James Adams recounts his experience in using this puzzle on a flyer to advertise his course in problem solving. He received one of his flyers back with a pencil stuck through it. The flyer was folded in such a way that all nine dots were on top of each other and the pencil, like an arrow, pierced through all of them.[4] A three line solution involves noticing that the dots have some non-zero width, and a one line solution involves adopting a 'global' problem space:

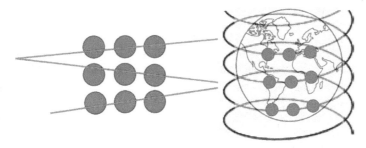

Thinking Outside the Box

A favorite one-line solution, however, was sent to James Adams by a five year old girl who managed to solve the problem within the confines of the square outlined by the nine dots using only one line!

Isolating the real problem involves correctly representing the given state and the goal state, adding constraints if necessary to sharpen the problem, and locating the problem in the correct problem space.

Avoid Premature Closure by Setting Idea Quotas

We often are saddled with mediocre solutions because we have a tendency to prematurely fix on one solution or one particular kind of solution. This tendency is known as *premature closure*. For example, in the previous problems many people prematurely fix on their own solutions rather than attempting to formulate the problem more precisely.

To avoid premature closure, it is useful to set an *idea quota*–decide in advance that you'll try to think up say, five alternatives before you choose the best one. It's not only the quantity of solutions that counts. Critical comparison among the alternative proposals often establishes constraints that may be useful to you in generating solutions of higher quality. Consider the following problem.

Flagging the Problem

It is 1970. The location is Berkeley, California. It is the day after the Kent State incident in Ohio in which several innocent students were shot and killed by national guardsmen during an anti-Vietnam war protest. A group of angry student protesters stormed the local McDonald's demanding that the flag be flown at half-mast in memory of the students killed at Kent State. The manager was sympathetic, especially in view of the size of the group, and he lowered the flag.

Not long after, the manager received a call from Ray Kroc, the ultraconservative owner of the McDonald's chain. Kroc demanded that the flag be raised back up immediately ("Or you've sold your last Big Mac!" he said).

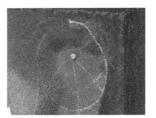

Bullet hole in Don Drumm's sculpture, Solar Totem #1, at Kent State

So the manager has a dilemma: risk the ire of the mob by raising the flag or risk his job by leaving it lowered.

Suppose you're the manager. Before you read on, try to come up with a completely satisfactory solution of your own.

This time try to think of at least five alternative solutions. Then choose the best among your proposals. Did you come up with a different solution? An improved solution? When I pose this problem in class, I often get answers such as:

- Tell the student protesters that you really need the job and that you'll lose your job if you comply with their wishes.
- Lie to Ray Kroc–tell him the flag is not at half mast, or that the students keep lowering the flag.
- Fly two flags–one at the top and one at half-mast.
- Try to get Ray Kroc to talk to the students and work things out.
- Offer the students free french fries.
- Tell Ray Kroc that the students will damage his building.
- Tell Ray Kroc off, fly the flag at half-mast, and join the student protesters.

Obviously not all these solutions are equally acceptable. But they give more options to choose from, and more importantly, they may suggest ways in which an initial solution might be improved. Even if you don't change your mind completely and adopt a totally different solution, if you're like most people, you'll find ways of improving your solution by incorporating the advantages of alternative proposals. Why does this happen?

Brainstorming is the process in which a group of individuals try to solve a problem by coming up with a lot of solutions off the top of their heads. It is important not to be too critical at the initial stages. The idea is to propose as many different solution paths as possible. A bad solution can always be rejected later, but good solutions are often overlooked because we tend prematurely to focus on the solutions that first come to mind. Some people think that brainstorming works because it prevents 'premature closure', the tendency to be satisfied with immediate and usually mediocre solutions.[5]

Superficially, premature closure does seem to give a good account of some failures of invention, as well as advice on how to do better. Edward de Bono, a researcher on creativity, uses the following analogy: "You cannot dig a hole in a different place by digging the same hole deeper."[6]

However, such talk of searching longer (or digging another hole) neglects something important: in the last analysis, problem solvers' evaluations of candidate solutions are responsible for leaving them with mediocre ones. Consider the McDonald's problem. The job was to devise an 'entirely satisfactory' solution. If later, after inventing a first solution, someone feels it is not entirely satisfactory, that means he or she wasn't critical enough in the first place. Of course, an entirely satisfactory solution may not exist–the problem may be to choose which solution is the most acceptable among the various imperfect alternatives.

The premature-closure concept leads us to think of premature closure as a matter of not searching long enough. But perhaps it's as much, or more, a matter of not requiring enough of the candidate solution we accept. Perhaps it's being sufficiently and insightfully critical that keeps the search moving forward rather than merely seeking a greater number of alternatives.

The devising and quick acceptance of mediocre solutions is not caused by short searches. Short searches are often the symptom of the failure to be critical. We either do not know, or know and do not maintain standards–whether these are explicit standards enforced by analysis or implicit ones derived from paired comparisons. The problem is not so much the lack of available solutions, but the acceptance of poor ones.

By the way, the solution that the manager from McDonald's came up with was probably an attempt to solve the wrong problem: how can I fly the flag at the top of the pole at half-mast? On the way out, the manager told the boy driving the delivery truck to back into the pole.

Working Forward and Backward

Sometimes you will get stuck trying to work from the given state to the goal state. If so, you might try working back from the goal state to the given state. Reverse gears and work backwards. You have probably found that it's often easier to solve a maze backwards rather than forwards.

The documentary *The Unknown Chaplin* [1983], shows a film sequence in which a strongman swings an ax down at Charlie. It just misses his head, and embeds itself in the ground in front of him dangerously close to his feet. Charlie nonchalantly steps over the ax, but while he's looking down at the ax, his hat falls off his head onto one of his feet. He looks at the hat, tilts his head, and with a deft kick, makes the hat miraculously fly off his foot right onto his head again. He then wanders off, hat on head, seemingly oblivious to the amazing feat of balance and physics he has just performed.

CHARLIE CHAPLIN (1889-1977)

How was it done? In order to achieve some of his startling cinematic stunts, Charlie Chaplin used to film the sequences backwards. When actually filmed, Chaplin walks backwards to the ax which is already embedded in the ground. He then tilts his head slightly forward, and lets his hat fall onto his foot. When shown in reverse, he appears to flip his hat from his foot to the top of his head, making audiences of the time wonder how many takes he did to achieve such dexterity and coordination!

Here's a game that illustrates the usefulness of working backwards.

A game is played with 17 markers. Two players take turns taking 1, 2, 3, or 4 markers at a time. The player to take the last marker wins. Which player should win–the first or second player?

Here's a sample game:

Round of Play	First Player	Second Player
First	Takes 3 ●●●	Takes 4 ●●●●
Second	Takes 1 ●	Takes 4 ●●●●
Third	Takes 2 ● ●	Takes 3 ●●● Wins

After playing the game a few times, you'll notice that if you've left your opponent with certain number of coins to choose from you have a forced win. Now add one coin at a time, and work backwards to see which player should win this game if both sides make the best possible moves.

A logical variation of this strategy of working backwards is the form of reasoning known as *reductio ad absurdum* (Latin for 'reduction to absurdity'), which can be seen as reasoning your way backwards into the truth. *Reductio ad absurdum* has the following form: Assuming some proposition is true, we deduce from it an impossible or absurd conclusion. We may then infer that the original assumption cannot be true.

The following puzzle, for example, can easily be solved by *reductio ad absurdum*.

Alonzo, his sister, his son and his daughter all play chess. The best player's twin and the worst player are of the opposite sex. The best player and the worst player are the same age. Who's the best player?

You can reason 'backwards' using *reductio ad absurdum*. First let's assume for the moment that Alonzo is the best player. On that assumption, the best player's twin would have to be Alonzo's sister. Now since the best player's twin (Alonzo's sister) and the worst player are of the opposite sex, the worst player must be a male. Since we're assuming Alonzo is the best player, the worst player would then have to be Alonzo's son. Now we're also told that the best player (Alonzo) and the worst player (Alonzo's son) are the same age. But this can't be. We've deduced an impossible state of affairs from our initial assumption.

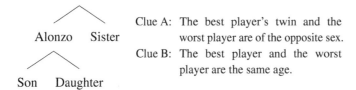

Clue A: The best player's twin and the worst player are of the opposite sex.
Clue B: The best player and the worst player are the same age.

Hypothetical Thinking and Reductio Ad Absurdum

1. *Assume*: Alonzo is the best player.
2. Alonzo's sister is the best player's twin.
3. The worst player is male.
4. Alonzo's son is the worst player.
5. Alonzo is the same age as his son,
6. So, Alonzo is not the best player.

Reducing our initial assumption to an absurdity, we can conclude that, Alonzo, at least, can't be the best player. By taking each possibility in turn, you'll eventually eliminate all the impossibilities and arrive at the solution to the puzzle.

Changing Your Conceptual Mode

Often we get blocked in our problem solving because we fail to adopt the best point of view from which to solve the problem or we fail to use the best problem solving language.

Efforts to design sewing machines got nowhere as long as the needle, with a point at one end and the eye at the other, had to pass completely through the fabric and be returned, but in 1846 Elias Howe changed his perspective. He put the eye of the needle in its point and solved the problem.

If you've ever grown your own tomatoes, you might wonder why there are so many thick-skinned tomatoes at the grocery store. The first mechanical harvesting machines handled the produce rather roughly; for fragile crops such as tomatoes, there was a lot of damaged produce. There was work done to make the machines less rough, but then researchers at U. C. Davis attacked the problem from another perspective. They worked on breeding tougher tomatoes.

When you can't solve a problem, it is often useful to change your point of view. Perhaps instead of changing the given state into the goal state, you can try to making the goal state closer to the given state.

Changing your point of view often has the effect of forcing you to step back and view the forest. Consider the following puzzle.

> You are working with a power saw and wish to cut a wooden cube 3 inches on a side into 27 1-inch cubes. You can do this by making six cuts through the cube, keeping the pieces together in the cube shape. Can you reduce the number of necessary cuts by rearranging the pieces after each cut?

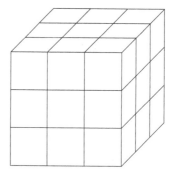

Can you cut the 27 cubes in less than 6 cuts?

After trying to solve the problem from your external point of view, try taking the point of view of the *inside cube*.

It is often useful to consider a variety of problem solving languages. Suppose your most important relationship is in trouble. One way to attack the problem would be to go to the library and read a book entitled *What's Wrong with Your Partner or Communication in a Relationship*.

This would be using a verbal problem solving language. Another approach would be to use your imagination. Imagine what it would be like if your relationship were working well.

What kinds of things would you and your partner be doing? What would you be saying to one another and how would you be saying it? What kinds of things would you be doing?

One way out of your present problems could be to start doing some of those things instead of what you're doing now. There are various problem solving languages you can use to formulate the problem and search for solutions.

As a child Einstein wondered, "What could I see in a hand held mirror if I were running at the speed of light?" In one of the most celebrated of Einstein's *Gedanken* (thought) experiments, Einstein visualized himself as a passenger riding on a ray of light and holding a mirror in front of him. He realized that he would see no image of himself, since he and the mirror were already moving at the speed of light and his own image could therefore never reach the mirror. Yet a stationary observer, also holding a mirror and seeing Einstein whiz by, would be able to catch Einstein's image in his own mirror. From this a fanciful visualization of physical events, Einstein gained the insight that led him to deduce his theory of relativity.

ALBERT EINSTEIN

Consider a contemporary example about using the best problem solving language. Chaos theory is a recent development in mathematics, its history going back to the work of Poincaré the great French mathematician. In 1887 King Oscar II of Sweden posed the question, "Is the universe stable?" and offered the equivalent of a Nobel prize for the best solution. Two years later Poincaré was awarded the prize for his celebrated work on the 'three body problem'.

Poincaré showed that even a system comprised of only the sun, the earth, and the moon but governed by Newton's law of gravity could generate dynamic behavior of such incalculable complexity that prediction, for all practical purposes, would be impossible. Although Poincaré had the mathematical insight, he did not have the technology to peer more deeply into the patterns within chaos. It was not until the advent of the computer that mathematicians discovered that deep within chaos there are hidden patterns known as *fractals*– intriguing objects which exhibit infinitely complex self-affinity at increasing powers of magnification.

The computer, with its capacity for handling complexity, provided the ideal problem solving tool to study chaos. Computers produces knowledge differently from traditional analytic instruments like the telescope or microscope. Scientists use telescopes and microscopes to analyze complex phenomena by reducing them to their simplest parts.

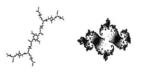

Fractals

The computer, on the other hand, simulates complexity. The fractal world was previously inaccessible not because it was too small or too far away, but because it was too complex for the unaided human mind.

Divide and Conquer

If you can't solve a complex problem, it is sometimes helpful to break the problem into simpler parts that you can solve. After you get an understanding of the overall picture, break the problem down into smaller problems and set subgoals before getting immersed in details. Turn means into ends. This heuristic is also known as *subgoaling*.

A classic example of subgoaling is the puzzle of the Towers of Brahma, more commonly known as the Towers of Hanoi, invented by mathematician Édouard Lucas, described by Henri de Parville in *La Nature*[7]:

> In the great temple at Benares beneath the dome which marks the center of the world, rests a brass plate in which are fixed three diamond needles, each a cubit high and as thick as the body of a bee. On one of these needles, at the creation, God placed sixty-four disks of pure gold, the largest disk resting on the brass plate and the others getting smaller and smaller up to the top one. This is the Tower of Brahma. Day and night unceasingly, the priests transfer the disks from one diamond needle to another according to the fixed and immutable laws of Brahma, which require that the priest on duty must not move more than one disk at a time and that he must place this disk on a needle so that there is no smaller disk below it. When all the sixty-four disks shall have been thus transferred from the needle on which at creation God placed them to one of the other needles, tower, temple and Brahmins alike will crumble into dust, and with a thunderclap the world will vanish.

Consider a simpler problem involving five disks. Reassemble the five disks on peg number two, in the same order, in the smallest possible number of moves. You may move only one disk at a time from one peg to another, and at no time may a larger disk be placed on top of a smaller disk.

To solve this problem it is useful to reduce it to a simpler problem you can solve. You can solve the problem of five disks if you can solve the problem of four disks. The problem of four disks can, in turn, be reduced to the problem of moving three disks. This last problem can be reduced to the trivial problem of moving two disks to another peg.

Towers of Hanoi Problem

Develop Schemata for Improvisational Fluency

Beginning in the 1930s Albert B. Lord researched the oral tradition in Yugoslavia and recorded his findings in *The Singer of Tales*[8]. The poets Lord studied performed at a rate of five to ten pentameter lines per minute, which is ten to twenty times faster than the rate at which the fastest poets could compose. Yet the oral poet does not repeat a memorized performance. She genuinely improvises. The poet will not sing or tell a tale exactly the same way on any two occasions.

Lord discovered that the oral poet relies on structured, prefabricated units–*schemata*. For example, the poet might have a stock of lines to describe the maiden's loveliness or the hero's armor. For the sake of flexibility, these units may have a variable element. For example, using the structure of part of the opening verse of *The Iliad* of Homer:

> *Sing, goddess, the anger of Peleus' son Achilleus and its devastation, which put pains thousandfold upon the Achaians...*

Try improvising a verse using the structure yourself.

Schemata are mental structures that facilitate our acting fluidly and effectively by allowing us to anticipate the organization of what we will perceive or perform. A gymnast with a repertoire of moves or physical schemata can learn new routines more easily than a novice and can create new routines using her repertoire as a basis.

So developing and maintaining schemata is a way of organizing knowledge in an effective way. Almost any activity requires some repertoire of schemata. New skills are learned through a mix of acquiring new schemata and adapting old ones, where the adaptation itself can be an occasion for creatively problem solving. Researchers have found, however, that such schemata, while leading to improvisational fluency, are no substitute for critical and careful composition. Mozart was admired in high society for his amazing fluency as an improviser,

Fuga from quartet in D minor, K. 173.
—Mozart

yet his notebooks reveal that when it came to producing his masterpieces, he was a painfully slow and meticulous composer. Mozart's repertoire of musical sketches—patterns and schemata—were a part of his musical genius but were no more important than the development of his sensitivites through exposure to music over his lifetime.

A *haiku* poem is the shortest form of Japanese poetry. The haiku is typically defined as a poem of seventeen syllables (*onji*), in three lines of five, seven, and five syllables. (Actually, the Japanese word onji does not mean 'syllable' but rather 'sound syllable' which is shorter and simpler than most syllables.) The form of seventeen onji derives from the traditional view that this is the optimum length of human speech to be delivered clearly and coherently in one breath.

The following haiku is by Basho, who is regarded as one of the greatest of the haiku poets :

Furuike ya, kawazu tobikomu, mizu no oto

whose sense is captured by the translation

Breaking the silence
Of an ancient pond,
A frog jumped into the water—
A deep resonance, [9]

and whose sound is perhaps better captured by the translation

Old pond...
a frog leaps in
water's sound. [10]

Notice how this haiku vividly captures the 'suchness' of experience–like a stone dropped into a quiet stillness of the soul.

To illustrate how one might obtain improvisational fluency by having schemata, as well the artistic limits of such schemata, consider the following amusing attempt to generate haiku poetry [11]. You begin by choosing one word from each of the following nine lists.

1	2	3	4	5	6	7	8	9
White	Buds	See	Snow	Trees	Spring	Bang	Sun	Flit
Blue	Twigs	Trace	Tall	Peaks	Full	Hush	Moon	Fled
Red	Leaves	Glimpse	Pale	Hills	Cold	Swish	Star	Dimmed
Black	Hills	Flash	Dark	Streams	Heat	Pfft	Cloud	Cracked
Gray	Peaks	Smell	Faint	Birds	Sun	Whizz	Storm	Passed
Green	Snow	Taste	White	Specks	Shade	Flick	Streak	Shrunk
Brown	Ice	Hear	Clear	Arcs	Dawn	Shoo	Tree	Smashed
Bright	Sun	Seize	Red	Grass	Dusk	Grrr	Flower	Blown
Pure	Rain		Blue	Stems	Day	Whirr	Bud	Sprung
Curved	Cloud		Green	Sheep	Night	Look	Leaf	Crashed
Crowned	Sky		Grey	Cows	Mist	Crash	Child	Gone
Starred	Dawn		Black	Deer	Woods		Crane	Fogged
	Dusk		Round	Stars	Hills		Bird	Burst
	Mist		Square	Cloud	Pools		Plane	
	Fog		Straight	Flowers			Moth	
	Spring		Curved	Buds				
	Heat		Slim	Leaves				
	Cold		Fat	Trees				
			Burst	Pools				
			Thin	Drops				
			Bright	Woods				
				Hills				

We've chosen the entries marked in red in the columns above. Choose your own list of nine words and plug them into the appropriate slots using the following schematic haiku:

ALL (1) IN THE (2)

(3) (4) (5) IN THE (6)

(7)! THE (8) HAS (9).

Using our own list of nine words, we arrive at the following:

> All pure in the snow
> Smell faint leaves in the mist
> Hush! The Moon has gone.

What did you get? Was it junk or a jewel? Could it be passable haiku with suitable revisions or was it complete nonsense?

We cannot simply define haiku poetry from the standpoint of syllabic structure alone. Haiku has grown out of a long historical and aesthetic process. Two features, for example, were at one time considered to be essential to starting a haiku.

First, a haiku poem should have reference to the season in which it is written. Secondly, there should be a breaking word (*kireji*)—a short emotionally charged word which, by arresting the flow of poetic statement for a moment, gives extra strength and dignity. The above schema tries to build such features into the schematic haiku.

Less structural and more dependent on aesthetic sensitivity are the subtle differences among the four traditional moods of the haiku.

The moods of *sabi, wabi, aware,* and *yugen* can best be defined by example.

Sabi: beauty with a sense of loneliness in time...

> *With the evening breeze,*
> *The water laps against*
> *The heron's legs.*

Wabi: the unexpected recognition of the faithful 'suchness' of very ordinary things...

> *A brushwood gate,*
> *And for a lock–This snail.*

Aware: the echo of what has passed and of what was loved...

> *The stream hides itself*
> *In the grasses*
> *Of departing autumn.*

Yugen: mystery, elegance, depth...

> *The skylark*
> *Its voice alone fell,*
> *Leaving nothing behind.*

Having the above schema may enable you to compose 'haiku' with great fluency, but to compose a good haiku, having experience with the the art form is absolutely necessary.

The following exercise illustrates how critical aesthetic judgment can be sharpened through the process of paired comparisons. Below are two sets of four haiku poems and each set contains one of each of the above four moods.

List A

A trout leaps,
   clouds are moving
      in the bed of the stream

Winter desolation
   in the rain-water tub,
      sparrows are walking

Sleet falling,
   fathomless, infinite
      loneliness

The evening haze,
   thinking of past things,
      how far-off they are

List B

Leaves falling
   lie on one another
      the rain beats on the rain

The woodpecker
   ...keeps on in the same place
      day is closing.

In the dark forest
   ...a berry drops
      the sound of water

In the dense midst
   what is being shouted
      between the hill and the boat?

Place a piece of paper over Lists A and B. Reveal one at a time a haiku from List A. After you reveal the first, try to identify its mood–choosing one of the four moods sabi, wabi, aware, or yugen. Then reveal the first two haiku and try to assign moods to the pair, assuming that each of the our moods is used exactly once. Then reveal the top three haiku, assigning to each one of moods. Finally, reveal all four with your best pairing of haiku and moods. Keep track of your series of revised guesses in a chart like this:

Reveal 1	Reveal 2	Reveal 3	Reveal 4
Sabi	Sabi	Yugen	Yugen
	Aware	Aware	Wabi
		Sabi	Sabi
			Aware

Now repeat the process for List B. Did you improve on your second attempt? Do you think your choices were correct earlier in the process of comparisons? Did the process of paired comparisons begin to train your aesthetic sense of haiku or not? If so, how? If not, why not?

The point is that schemata such as the one above can lead to improvisational fluency, but truly creative and artistic work also requires critical sensitivities forged by experience, in this case being steeped in the traditions and history of the haiku art form. There's no substitute for experience. By the way, the answers to List B are: *yugen, sabi, wabi,* and *aware* (listed here in reverse order!)

Let the Problem Incubate

If at first you don't succeed, it's not always a good idea to try, try again. After you have worked on the problem long enough to have it firmly set in your mind, it is often helpful to let the problem incubate. Sometimes the solution will pop into your mind unexpectedly. The mathematician Henri Poincaré described his own process of mathematical creativity as follows:

Henri Poincaré (1854-1912)

> It is time to penetrate deeper and to see what goes on in the very soul of the mathematician. For this, I believe, I can do best by recalling memories of my own... For fifteen days I strove to prove that there could not be any functions like those I have since called Fuchsian functions. I was then very ignorant; every day I seated myself at my work table, stayed an hour or two, tried a great number of combinations and reached no results. One evening, contrary to my custom, I drank black coffee and could not sleep. Ideas rose in crowds; I felt them collide until pairs interlocked, so to speak, making a stable combination. By the next morning I had established the existence of a class of Fuchsian functions... Just at this time I left Caen, where I was then living, to go on a geologic excursion under the auspices of the school of mines. The changes of travel made me forget my mathematical work. Having reached Coutances, we entered an omnibus to go some place or other. At the moment when I put my foot on the step the idea came to me, without anything in my former thoughts seeming to have paved the way for it, that the transformations I had used to define the Fuchsian functions were identical to those of non-Euclidean geometry... On my return to Caen, for conscience's sake I verified the result at my leisure. [12]

Poincaré goes on to describe how once again this characteristic cycle of intense intellectual preparation, incubation, sudden illumination, and careful verification lead to further mathematical creativity. Poincaré then went beyond *describing* a phenomenon and offered his own *explanation* of it:

> Most striking at first is this appearance of sudden illumination, a manifest sign of long, unconscious work. The role of this unconscious work in mathematical invention appears to me incontestable, and traces of it would be found in other cases where it is less evident. Often when one works at a hard question, nothing good is accomplished at the first attack. Then one takes a rest, longer or shorter, and sits down anew to the work. During the first half-hour, as before, nothing is found, and then all of a sudden the decisive idea presents itself to the mind...

If incubation occurs when time away from a problem helps to solve it, then there would be many explanations other than Poincaré's hypothesis of extended 'unconscious thinking'. Can you think of other possible explanations for the effectiveness of incubation?

As a classic example of incubation, we'll conclude this chapter with the story of Archimedes and his bath. In so doing we'll also tell the story of the Greek roots for the term 'heuristics'. Hiero, the new ruler of Syracuse, had commissioned a golden crown for the occasion. However, an accusation was made that some of the gold had been stolen and the weight made up with mere silver.

Eureka!

How could Hiero be sure? To melt the crown down would be to destroy all the work the artisans had done. Hiero sought an answer from the renowned mathematician Archimedes. Archimedes was aware that gold was 'heavier' than silver, i.e., that silver has more volume than gold for the same weight. He therefore deduced that a crown made of silver and gold would be slightly larger than one of the same weight made of pure gold. But the crown was so irregularly shaped that there was no ready way to compare it by volume with the same weight of gold.

Worrying about this problem, Archimedes let the problem incubate and decided to go to the public bath to soak and to think about the problem afresh. On getting into the tub, Archimedes noticed that the deeper he settled, the more the water flowed over the edge of the bath.

All at once he had it. His body was displacing an equal volume of water from the tub. Similarly, the volume of the crown, even if irregularly shaped, could be measured by immersing it in a full tub of water and measuring the overflow. Archimedes, so the story goes, sprang from the tub and ran naked through the streets of Syracuse shouting "Eureka!" which means in Greek, "I have found it!" It is from this Greek expression that we derive our term 'heuristic.'

Summary of Concepts

A PROBLEM exists when what is (the given state) differs from what is wanted (the goal state).

A SOLUTION PATH is a series of operations you can perform to transform the given state into the goal state.

There are often CONSTRAINTS on what can count as an acceptable operation in the solution path.

The solution takes place in a context known as the PROBLEM SPACE.

PREMATURE CLOSURE occurs when one runs with a solution that comes to mind and fails to search long enough or critically enough.

INVENTION is the process of searching for solution paths through a space of alternatives.

SCHEMATA are mental structures that allow us to perceive or perform effectively by anticipating the organization of what we apprehend or do, so we needn't function as much from scratch.

HEURISTICS are rules of thumb, strategies, tricks, simplifications, or any other kinds of devices which limit the search for solutions in a space of alternatives. Some of the heuristics we've discussed are:

* isolate the real PROBLEM
* avoid PREMATURE CLOSURE by setting IDEA QUOTAS
* work FORWARDS AND BACKWARDS
* change your CONCEPTUAL MODEL – your point of view or problem solving language.
* divide and conquer by SUBGOALING
* develop SCHEMATA AND CRITICAL EXPERIENCE
* let the problem INCUBATE

Exercises

Group I

1. **What's the Problem?**

   Quickly jot down your answers to the following problems.

   (A) A flag pole is 15 ft. tall, and its shadow is 45 ft. longer. How many times is the shadow longer than the flagpole?
   (B) A brick weighs ten and one-half lbs. plus half its own weight. What's its total weight?

(C) In your head, divide 30 by one-half and add ten. What's the answer?

(D) A bottle and a cork together cost a $1.10. The bottle costs a $1 more than the cork. How much does the cork cost?

(E) Teresa sold her computer to Saul for $1000. After using the computer for a few days, Saul discovered that its chip was defective so he sold it back to Teresa for $800. The next day Teresa sold the computer to Sui Ling for $900. What is Teresa's total profit?

Now go back and carefully read each of the problems. Did you identify the given and goal states correctly?

2. *Picking the Problem Space*

Take six toothpicks and form exactly four equilateral triangles. (Of course, you may not break or bend the toothpicks. I know of several clever solutions–all of which involve casting off unwarranted assumptions.)

Group II

3. *Weighing Alternatives*

You have a pile of 24 coins. One of the coins is a counterfeit.

(A) You are told that twenty-three of these coins have the same weight, but one is heavier than the rest. You also have a balance scale to compare the weight of any two sets of coins. How can you determine which is the heavier coin with a smallest number of weighing?

(B) This time you are told that one of the coins is either heavier or lighter than the rest, but that you don't know which. Again you have only a balance scale to compare the weight of any two sets of coins. Now what is the smallest number of weighing needed to guarantee that you have found the counterfeit coin?

4. *Inverting*

You have six arrows in the row arranged as follows:

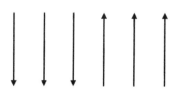

The goal is to transform the arrows into an alternating sequence with the left-most arrow up, the next pointing down, the next pointing up, and so forth. However, you're not allowed to turn just one arrow over at a time. Instead you must simultaneously invert any two adjacent arrows. What's the smallest number of moves required to obtain an alternating sequence?

5. Back and Forth

You have a jar that will hold exactly 7 liters of water and another that will hold 3 liters of water. You have a river but no other containers. Describe a sequence of filling and emptying the water jars that ends with having exactly 5 liters of water.

6. Reverse Doubling

Three people play a game in which one person loses and two people win each game. The loser must double the amount of money each of the two players has at the time. The three players play three games, at the end of which each player has lost one game and each has $8. What was the original stake of each player?

7. Word Golf

Word golf was invented by Lewis Carroll, the author of *Alice in Wonderland*. (He called it 'doublets'.) In word golf you begin with a word and then replace one letter of the word at a time. Each replacement must result in another word without rearranging the letters. For example, you may change HATE to LOVE as follows:

HATE ⇒ LOVE   PAR 4

0.	HATE
1.	HAVE
2.	RAVE
3.	ROVE
4.	LOVE

If we had used the unusual English word 'LAVE' this derivation may be shortened to only three word shifts: HATE, HAVE, LAVE, LOVE, which is better than 'par for the course.' Here are a number of word golf courses for you to play. Included in parentheses is an indication of what is 'par for the course.' (It is possible to beat the par for the course).

(A) Evolve APE into MAN (par 5).
(B) Change BEER into WINE (par 6).
(C) Make the OLD NEW again (par 8).
(D) SUMMER passes into WINTER (par 8).
(E) Accelerate from SLOW to FAST (par 8).
(F) Transition from BIRTH to DEATH (par 11).
(G) Change from DRUNK to SOBER (par 13).
(H) Bring ORDER out of CHAOS (par 13).

8. Northcott's Nim[13]

Consider Northcott's Game, which is played on a chess or checker board. The two players each place one of their eight pieces on each file (vertical column). Moves consist of sliding a piece either up or down on the file. (You cannot jump over the opponent's piece, move to another file, or move two pieces at once.) The winner is the last player to make a move. Assuming it is white's turn, what is the winning move?

9. 21 and 2001

If both players play their best strategies, who should win the following games–the first or second player?

(A) In the game of 21, two players take turns, beginning at zero, adding either 1, 2, 3 or 4. The first player to reach 21 wins.

(B) In the game of 2001 the players begin with the number 2001, from which each player subtracts a number from 1 to 99 on each turn. The player to reach 0 first wins.

10. Subtotals

How many different downward paths from A to B are in the grid?

11. Double Crossing

One morning at sunrise a Buddhist monk began to climb a mountain along a narrow path that wound around it. He climbed at varying rates of speed and stopped from time to time to rest. He reached the top of the mountain, where there was a temple, at sunset. He remained at the temple to meditate for several days. Then at sunrise one morning, he started down the same path, again walking at varying rates of speed, though his average speed of descent was greater than his average speed of ascent. Must there be a spot along the path that he will occupy on both trips at exactly the same time of day?

12. Easy as Pi

Imagine that you are on a perfectly round sphere the size of the earth. A steel band is stretched smoothly around the equator. One yard of steel is added to this band so that it is raised off the equator by the same distance all around. How high will this lift the band? Which of the following (A-E) is the closest to being the biggest object you can slip under the band?

Use your imagination first, then use a little algebra.

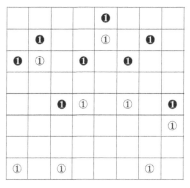

NORTHCOTT'S NIM

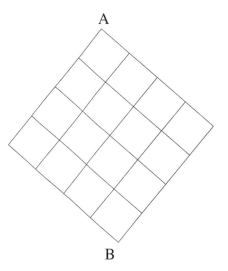

(A) you can slip a playing card underneath;
(B) you can slip your hand underneath;
(C) you can slip an orange underneath;
(D) you can slip a volleyball underneath;
(E) there is not enough information to solve the problem.

The circumference of a circle is $2\pi R$, where R is the length of the radius of the circle.

```
        X X 8 X X
X X X )X X X X X X X X X
        X X X
         X X X X
          X X X
           X X X X
            X X X X
```

13. *The Lonesome 8 Problem*

This is a classic cryptoarithmetic problem. Substitute digits between 0 and 9 inclusively for each X so that the result is a correctly solved problem in long division. (You may assume that the initial digits in a string are never zero.)

14. *Dividing the Loaves*

According to a legend, two travelers stopped at a small town for a meal. One had five loaves of bread and the other had three loaves. Before they began eating they were joined by a hungry stranger who asked for food and offered to pay for what he ate. They agreed to divide their bread with him so he sat down and ate. They divided the bread equally, and when the meal was over, the stranger laid down eight coins of equal value in payment for the food, and then he went on his way.

The traveler who had had five loaves took five of the coins as his share, and the other was left with three. The second was not pleased with this division and argued that he should receive half of the coins. The two could not agree on this, and they argued so vehemently that finally they had to take their case to a judge.

The judge listened to the story of what happened, and then he said, "The man who had five loaves should receive seven of the coins, and the man who had three loaves should receive only one."

Which is the correct division of coins?

15. *Reflections*

The following problems involve imagining mirror images.

(A) Theoretically, what is the minimum length of a plane mirror in order for you to see a full view of yourself? (You might want to experiment with an actual mirror.)
   1. 1/4 your height;
   2. 1/2 your height;
   3. 3/4 your height;
   4. your full height;
   5. it depends on your distance from the mirror.

(B) Without looking in a mirror, write down your name on a piece of paper so that it will appear to be correctly written to you when you are looking at the paper's reflection in the mirror.

16. *QWERTY*

The following is a template of keys for the standard letters and numbers on a keyboard. Although you have used a keyboard many times, you may not recall the keys for all the letters, numbers and symbols. Without looking at a keyboard, try to fill in the following template. What did you find is the most appropriate problem solving language?

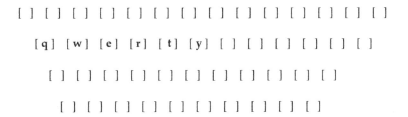

17. *Toe-tac-tic*

Two players are playing *reverse* tic-tac-toe. The object of the game is to avoid getting three in a row. The first player has the Os and the second player has the Xs. The players take turns entering their Xs or Os on the standard 3 x 3 tic-tac-toe board. In contrast to tic-tac-toe, in toe-tac-tic it is very difficult for the first player to obtain a draw unless the player knows the strategy. What strategy can the first player use to insure a draw at toe-tac-tic?

18. *Dividing the Camels*

Many years ago an Egyptian died and left his estate, which consisted of seventeen camels, to his three sons. The eldest was left half of the camels, the second son was left a third of the camels, and the youngest son was to receive one-ninth of the camels, according to the will. The sons and their legal advisers could figure no way to make the distribution without cutting up a camel. So they went to the Pharaoh for advice. The Pharaoh, after listening to the problem, thought for a moment and then announced a solution which had not occurred to anyone. What do you think it was? (For a generalization of this problem, see Ian Stewart, "The Riddle of the Vanishing Camel", *Scientific American*, June 1992, together with a correction in the next issue.)

19. *Bicycles and the Bee*

Two bicyclists are 20 miles apart and begin racing directly toward each other. The instant they start, a bee on the wheel of one bicycle starts flying straight toward the other cyclist. As soon as it reaches the other bicycle's wheel it turns and starts back. (You may assume the turning time is negligible.) The bee flies back and forth in this way, from wheel to wheel until the two bicycles meet. Each bicycle had a constant speed of 10 miles an hour, and the bee flies at a constant speed of 15 miles an hour. How far did the bee fly before it was squashed?

20. *Frame of Reference*

You are standing by the side of a river which is flowing past you at the rate of 5 miles an hour (from left to right). You spot a raft 1 mile upstream on which there are two children adrift. Then you spot their parents 1 mile downstream paddling upstream to save them. The parents usually paddle in still water at a rate of 4 miles an hour. If the parents paddle at their usual strength against the current towards the children, how long will it take for them to reach the children?

21. *The Walking-Fly*

A living room is 30 feet long, 12 feet high, and 12 feet wide. There's a lollipop smudge on the wall of the living room 1 foot from the floor and 6 feet from each corner. A fly with a broken wing is standing on the opposite wall 1 foot from the ceiling and 6 feet from each corner. What's the shortest distance the fly can walk to get from where he is to the lollipop smudge on the opposite wall?

22. *Ladybugs Cubed*

A pair of ladybugs crawl on a paths across two identical square tiles. The ladybug paths on the tiles are indicated by the blue line and the broken purple line. The ladybugs reach the dividing line two tiles simultaneously. At that moment, the second tile is tilted up along the dividing line so that the tiles make a 90 degree angle. What angles do the ladybug paths make in space?

23. *Metaphors*

Metaphor is not merely ornamental writing, a beautifying cosmetic laid on the surface of thought. Metaphor is also a process by which we see more, and more deeply. By finding resemblances between remote objects or ideas, metaphorical-analogical thinking opens new pathways of thought.

Consider the metaphors contained in the famous soliloquy from Shakespeare's *MacBeth*, Act V, Sc. 5:

To-morrow, and to-morrow, and to-morrow,
Creeps in this petty pace from day to day,
To the last syllable of recorded time;
And all our yesterdays have lighted fools
The way to dusty death. Out, out brief candle!
Life's but a walking shadow, a poor player
That struts and frets his hour upon the stage
And then is heard no more; it is a tale
Told by an idiot, full of sound and fury,
Signifying nothing.

What's the literal and non-metaphorical meaning of this passage? That everyone dies after a brief life that has no meaning. But notice how flat, how thin, how impoverished the paraphrase is. And that is not merely because it is prosaic and flat, but also because it conveys so much less information and so much less wisdom than the metaphors. How can this be?

Why does metaphor accomplish this? George Lakoff and Mark Johnson in *Metaphors We Live By* [14] offer an intriguing (but metaphorical) explanation: all the propositions connected to one half of the metaphor—all its 'entailments'—become transferred to the other half of the metaphor, adding a great deal of conceptual content to it. Consider, for example, the metaphor,

Love is a collaborative work of art.

It has the following entailments:

> Love is work.
> Love requires compromise.
> Love is an aesthetic experience.
> Love requires discipline.
> Love is creative.
> Love cannot be achieved by formulas.

Each of these may, in turn, have other entailments, all of which, subsumed by the metaphor, provide new meaning to the concept of love. "What we experience with such a metaphor," write Lakoff and Johnson, "is a kind of reverberation down through the network of entailments that awakens and connect the memories of our past love experiences and serves as a possible guide for future ones."

Using this model of metaphor, try to explain some meanings to be found in the following metaphors found in Shakespeare.

(A) *The quality of mercy is not strain'd*
It droppeth as the gentle rain from heaven.
Upon the place beneath.
  —*The Merchant of Venice*, Act IV, Sc. 1

(B) *Grief fills the room up of my absent child.*
Lies in his bed, walks up and down with me.
Puts on his pretty looks, repeats his words,
Remembers me of all his gracious parts,
Stuffs out his vacant garments with his form.
  —*King John*, Act III, Sc. 1

(C) *He draweth out the thread of his verbosity*
finer than the staple of his argument.
  —*Love's Labour's Lost*, Act V, Sc. 1

Group III

24. *Convergence*

Once you recognize problem solving strategies that are potentially useful, it can be handy to abstract the general characteristics from the analogs to use in solving other problems. M.L. Gick and K.J. Holyoak, for example, represent the general characteristics of the Dunker stomach tumor problem as follows:[15]

Convergence Schema

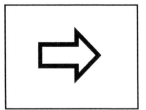 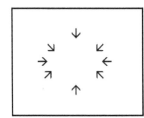

Invent stories for three different problems that can be solved using the convergence schema. Each of the stories should involve a problem which can be overcome with a strong force, but in which the force must be varied in strength and in which multiple pathways must be used to carry the force to the target. You might consider translating the spatial schema into a temporal one, or a two dimensional schema into a three-dimensional one.

25. *Nine Dot Problem*

A reason given for why people often fail to solve the nine dot puzzle is that they make unwarranted assumptions about the problem space. R.W. Weisberg and J.W. Alba tested the idea that removing the problem solver's fixation on the square boundary bounded by the dots would result in the solution.[16] In their experiment, one group (the control group) of subjects were given a maximum of 20 attempts to solve the problem. A second group were given 10 attempts and then were told that they would have to extend the lines outside the square to solve the problem. Their results showed that almost all of the hint subjects drew lines extending beyond the square, while none of the control group subjects did. Unfortunately, none of the subjects in either group solved the problem!

C.T. Lung and R.L. Dominowski studied the problem again.[17] This time both groups were given 40 attempts and the second group of subjects were given a hint and some time to practice with problems that involved drawing straight lines through dots. Here are some practice problems with solutions:

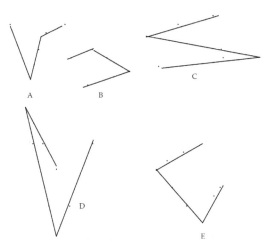

In contrast to Weisberg and Alba's results, Lung and Dominowski found that the nine-dot problem was solved by 60% of the students given the hint and practice time and only 9% of the control group. Design an experiment to test these findings.

26. *Pandemic Puzzles*

(A) In 2020 the population of the U.S. is estimated to be about 328.42 million. If 1 person infects 2 non-infected persons every day and once infected a person remains infected. At this rate, how long will it take for the entire U.S. population to be infected?

(B) In *Innumeracy: Mathematical Illiteracy and Its Consequences*, John Allen Paulos suggested a way of getting an intuitive sense of 1 million. How many days is 1 million seconds? It's about 11.5 days (more accurately, 11.57 days). Figure out your own way of communicating the difference between a debt of 1 million (10^6), 1 billion (10^9), and 1 trillion (10^{12}) dollars.

(C) The U.S. spends more on health care per capita than any other country in the world. Yet around 2010, U.S. life expectancy plateaued and in 2014 began reversing, dropping for three consecutive years–from 78.9 years in 2014, to 78.6 in 2017.[18] Why?

(D) In 2020 the average life expectancy United States for males is 76.3 and for females is 81.3 years old.[19] Why is it that the chances are better than ever that you'll live longer than the average?

(E) A class has 12 students who are distributed into Zoom breakout rooms containing either 3 or 4 students each. Every pair of students is together in a breakout room exactly once. What is the maximum number of days this can happen?[20]

3

Bridges to Problem Solving

...the path I followed will be of some help, perhaps...
—Leonhard Euler (1707-1783)

Our mental universe sometimes grows out of the problems we encounter in our everyday life much like the way our dreams arise out of the events of the day. If we have a sense of wonder, the world may serve as a bridge, a point of departure, suggesting problems that turn out to have wide application. One such problem is the celebrated puzzle of the seven bridges of Königsberg.

The Puzzle of the Königsberg Bridges

In the old town of Königsberg, Prussia, there were seven bridges that crossed the Pregel River. Here's a map of the seven bridges (marked by the lower case letters) joining the various land areas (marked by the capital letters.)

Leonhard Euler

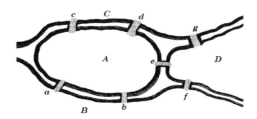

The 7 Bridges of Königsberg

For amusement on their evening walks, the townspeople of Königsberg posed for themselves the following puzzle: is it possible to plan a walk that crosses each of the seven bridges once but not more than once? Try it for yourself!

In 1735 this puzzle reached the Swiss mathematician Leonhard Euler (pronounced "oiler"). He soon recognized that there was an important mathematical principle contained in this seemingly trivial puzzle.

Although Euler realized that the particular problem of the seven bridges of Königsberg would be solved by carefully tabulating all possible paths and seeing whether any of them met the requirements, he quickly dismissed this method as too tedious and too restricted. Instead he posed a more general problem:

> Given any configuration for the river and the branches into which it may divide, as well as any number of bridges, determine whether or not it is possible to cross each bridge exactly once.

Euler decided to approach the original problem of the seven bridges of Königsberg by constructing a *model* or abstract representation of the configuration of bridges. To construct such a model, Euler needed to decide which features of the real world situation were relevant and which were not:

> The branch of geometry that deals with magnitudes has been zealously studied throughout the past, but there is another branch that has been almost unknown up to now; Leibniz spoke of it first, calling it the "geometry of position" (geometria situs). This branch of geometry deals with relations, dependent on position alone, and investigates the properties of position; it does not take magnitude into consideration, nor does it involve calculation with quantities.

Looking at the problem aside from its physical trappings, Euler noticed that it is irrelevant that the land areas have physical magnitude–so he represented them as points or vertices. Second, he noticed that only the relative positions of the bridges are relevant and these positions could be represented by arcs connecting the various points.

Abstracting from the real world situation, Euler obtained a connected graph. A problem equivalent to the original, but now stated in terms of the connected graph, is this: Can the connected graph be traced in one continuous line without retracing? (We'll call such a tracing an "Euler Path.")

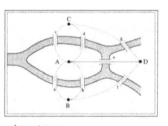

*Euler's Model of the
7 Bridges of Königsberg*

Real World Situation — Model

- The land areas can be shrunk to points or vertices: the actual area doesn't matter.
- Bridges can be represented by arcs connecting the points or vertices: the actual length doesn't matter.

Can you plan a walk that crosses each bridge once and only once?

Can you trace the graph without lifting your pen and without retracing a line?

Euler offered another algebraic way of representing the configuration of bridges.

> My entire method rests on the appropriate and convenient way in which I denote the crossing of bridges, in that I use capital letters A, B, C, D to designate the various land areas that are separated from one another by the river. Thus when a person goes from area A to area B across bridge a or b, I denote this crossing by the letters AB, the first of which designates the area whence he came, the second the area where he arrives after crossing the bridge. If the traveler then crosses from B over bridge f into D, this crossing is denoted by the letters BD; the two crossings AB and BD performed in succession I denote simply by the three letters ABD, since the middle letter designates the area into which the first crossing leads as well as the area out of which the second leads.

Using Euler's notation, state the problem of the Königsberg bridges.

Specializing & Generalizing

In order to get a feel for a problem, it is sometimes useful to examine some special cases and look for patterns. Try to find Euler paths for each of the following graphs. After you are done you should record your results in the chart below.

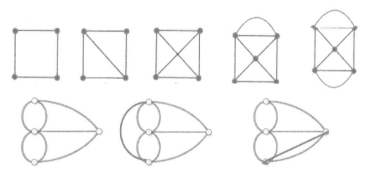
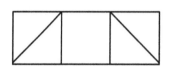

Perhaps you've noticed that sometimes your Euler path began and ended at the very same vertex and other times you began and ended at different vertices.

Let's introduce some terminology which will help us to express our insights. We'll call an Euler path *closed* if it begins and ends at the same vertex. An Euler path is *open* if its beginning and ending points are different.

It appears that the number of lines converging at a vertex or point is important. Let's call that number the degree of the vertex. It turns out that the actual number isn't as important as the parity of the point–whether the degree of the vertex is odd or even.

When looking for patterns, it is useful to organize your observations in a chart:

	Euler Path?	Open or Closed	# Odd Deg. Vertices
1			
2			
3			
4			
5			
6			
7			
8			
9			

Now look for a relationship between the first and third columns. When does a graph have an Euler path? Look at the second and third columns: if a graph has an Euler path, when it is closed? Fill in the blanks to complete the following conjecture:

> If a graph has an Euler path, then it has either ___ vertices of odd degree (in which case the path is CLOSED), or exactly ___ vertices of odd degree (in which case the path is ___).

But as one mathematician has put it, "the accumulation of examples is not mathematics any more than a dictionary is a novel."[1] Can you construct any counterexamples to your conjecture? No matter how certain we might be of our result on the basis of these examples, we cannot be certain of our result unless we have a proof. Let's try to see if we can prove our conjecture.

Euler's Theorem

Euler reasoned in a general way about the existence of Euler paths for a connected graph. Let's suppose a graph has an Euler path. Let's consider any vertex other than the beginning or ending vertices. For any such vertex, the Euler path goes into the vertex and leaves it. What does that do for the degree of the vertex–it adds a count of 2. If the Euler path passes through the vertex again, it adds a count of 2. Therefore, the degree of any such vertex is even.

Next consider a closed Euler path. In such case, we begin and end at the same vertex and so it will have a count of 2, plus 2 more any time the Euler path passes through the vertex. Therefore, all the vertices on a closed Euler path are of even degree.

Finally, consider an open Euler path with vertices A and B as the beginning and ending. The ending or beginning of the open Euler path contributes 1 to the degree of vertices A and B, and any time the path passes through one of these vertices, adds a count of 2. Therefore, the degrees of A and B must be odd. Therefore, if a graph has an open Euler path, then it will have exactly two vertices of odd degree, and these two vertices will be at the beginning and ending points of the Euler path. We have arrived at the following result:

> If a connected graph has an Euler path, then it will be either open or closed. If it is open, then all the vertices will be even except for two–the beginning and ending points. If the Euler path is closed, then all the vertices will be of even degree.

Euler actually proved something stronger:

> A connected path has an Euler path *if and only if* all of its vertices are of even degree (in which case it has a closed Euler path) or exactly two of its vertices are of odd degree (and those vertices will be the beginning and ending points of the Euler path.

In other words, not only did Euler prove that the two conditions were necessary for the existence of Euler paths, but that they were also sufficient. If a connected network has exactly two vertices of odd degree, then it will have an open Euler path; and if a connected network with all vertices of even degree, then it will have an closed Euler path. How might you go about proving this?

Test you understanding of Euler's theorem by applying it to the following network. Does it have an Euler path? If so, is the path open or closed? Show one example of an Euler path by tracing the network with one continuous line.

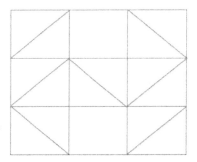

Trace an Euler Path

Notice that using the Euler theorem considerably shortened your search for a solution–improving your ability to solve the problem without the inefficiency of trial and error.

How does Euler's theorem apply to the original Königsberg bridge puzzle? Label the connected graph for the Königsberg bridges odd or even. How many vertices of odd degree does it have? Euler decisively solved the puzzle of the Königsberg bridges by proving that the failure of the inhabitants of Königsberg to solve the puzzle was no accident. The solution to their puzzle is that the puzzle is impossible to solve! Impossibility proofs are, in a way, more interesting than merely solving a puzzle by stumbling onto a solution. Proving that a problem is impossible to solve often requires us to conceptualize or model the given situation and then to give a general argument proving the impossibility of transforming the given state into the puzzle's goal state.

Some Advantages of Modeling

Euler's problem solving strategy was to abstract from the particular physical situation to obtain a model, and then to give a very general argument about such models. What's so useful about modeling?

Euler could have solved the problem of the Königsberg bridges in a very uninteresting way by simply enumerating all the possible paths and checking to see whether any of them satisfied the requirements of the problem. Such a process would be tedious (even performed by a computer) and probably would not, by itself, yield any general conceptual insights. Here, for example, is a chart using Euler's algebraic method for representing the problem. The chart begins to list the possible paths showing they all fail, beginning with a path from A to B. The hash mark (#) means the path cannot be continued.

A	B	A	C	A	D	B	#
A	B	A	C	A	D	C	#
A	B	A	C	D	A	C	#
A	B	A	C	D	B	#	
A	B	A	D	B	#		
A	B	A	D	C	A	C	#
⋮							

Euler's solution is more elegant than this "brute force" method. It enables us to solve problems with simple criteria. Does the graph have vertices all of which are of even degree? If so, you can begin an Euler path at any vertex and the Euler path will eventually close and end at that very same vertex. Does that graph have exactly two vertices of odd degree? If so, there is an Euler path that begins at one of those vertices and ends at the other. There are also some advantages of modeling over exhaustive searches.

First, modeling a problem forces us to make our assumptions explicit. We often unconsciously make assumptions about a given situation which may or may not be warranted. Here, for example, is the first line from a children's book:

> Lizzy heard the ICE-CREAM truck coming down the street. She remembered her BIRTHDAY MONEY and ran into the house.

What natural inferences do we draw from these scant two lines? How old do you think Lizzy is? What kind of neighborhood does she live in? Do you think that Lizzy likes ice-cream? Notice that many of the things we naturally assumed aren't actually explicitly stated. Instead we automatically supplied the missing information from our stockpile of world knowledge. And these assumptions can be dramatically changed by making two simple substitutions in the about two sentences:

> Lizzy heard the BUS coming down the street. She remembered her GUN and ran into the house.

Modeling often forces assumptions to the surface, where they can be either accepted or rejected. Here's another puzzle of that illustrates the same sort.

> A woman is running home. There is a man dressed in black watching and waiting for her. What is going on?

We are always making assumptions–that's a requirement for efficiently perceiving, thinking, and navigating our way through life. Yet, being aware of the fact that we are constantly making assumptions can be an occasion for creative problem solving. When we begin to critically evaluate whether our assumptions are valid we will adapt to situations with new solutions to old problems.

Euler's elegant theorem of connected graphs applies to diverse situations, from molecular connections to social networks, and can solve all of these applications in a simple mathematical way.

In constructing our model of the Königsberg bridges, we implicitly imposed a particular problem space. Without such a limitation, we might try to solve the problem by digging a tunnel, or perhaps by globe-trotting around the world in order to get to the other side of the Pregel river without crossing a bridge. Our modeling makes clear what assumptions we are making about the problem space showing that, sometimes, the problem space needs to be questioned or transcended.

Puzzle: Take six toothpicks and form four equilateral triangles.

There are several solutions that can be discovered by rolling the toothpicks around on the top of a table. See if you can find at least three different solutions. One of the most elegant solutions is to create four equilateral triangles by questioning the assumption that the puzzle solution needs to take place on the plane of the table top.

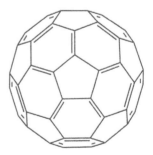

Buckminster Fuller (1895-1983) said he once discovered the idea behind his geodesic domes as a child with bad eyesight because when other children were building cubes with toothpicks and peas, he created a toothpick and pea structure by feeling the stability created by triangulation. He built a Platonic solid simpler than the cube (or hexahedron) out of equilateral triangles (the tetrahedron).

Having a model can improve the efficiency of our search for a solution. Due to the limits of short-term memory, we often search through the problem space in a serial fashion. When we reach a dead end, we back up and try again. Often we get tired or forget and go down the same road again and again. A good model makes it easier to keep track of our mistakes, allows us to make parallel searches, and even suggest ways to prune the tree of alternatives by exploiting symmetries or using other heuristics.

Consider the Towers of Hanoi puzzle invented by French mathematician Édouard Lucas in 1883. According to myth, there is a great Brahman temple at Benares. Beneath its dome, which marks the center of the world, there is a brass plate with three fixed diamond needles, a cubit high and as thick as the body of a bee. On one of these needles Brahma, the creator, has placed 64 golden disks, each with a decreasing diameter. These 64 disks must be transferred from the one needle to another by the priest of the temple according to the following immutable rules:

(1) Only one disk can be moved at a time.
(2) Each move consists of taking the top most disk from one of the stacks and placing it on top of another stack or on an empty needle.
(3) No larger disk may be placed on top of a smaller disk.

According to the legend, when the last move of the puzzle is completed, the temple and the entire world will crumble into dust with a thunderclap and disappear.

There is no need to worry. Assuming that the priests are able to move the disks at a rate of one per second, the smallest number of moves would take $2^{64} - 1$ seconds or about 585 billion years. Given that the age of the universe is roughly 14 billion years, this is approximately 42 times the current age of the universe. So, there is no question about solving the puzzle by moving the 64 disks one at a time.

Instead, we can solve a simpler version of the problem and look for patterns. It is simple to see that it takes a minimum of 3 moves to transfer the disks to another needle or peg. What is the minimum number of moves for 3 disks? Well, it takes 3 moves to move the top two disks to another peg, another move to transfer the largest disk to the empty peg, and then 3 more moves to transfer the tower of two on top of largest disk. This is a total of $3 + 1 + 3 = 7$ moves. To move 4 disks, we can use what we've already figured out. It takes 7 moves to move the top 3 disks to another peg, an 8th move to transfer the largest disk, and then 7 more moves to transfer the tower of 3 disks to the top of the largest disk, for a total of $7 + 1 + 7 = 15$ moves. So, 2 disks takes 3 moves, 3 disks takes 7 moves, 4 disks takes 15 moves, and, in general, if there are n disks, the minimum number of moves is 2^n-1.

There are other geometrical ways of keeping track of the moves. If you arrange the three needles or pegs as a triangle, you can see the solution to the puzzle by looking down at the triangle. Color the disks red and blue. It turns out that a minimal transfer involves rotating each of the red disks clockwise and rotating each of the blue disks counterclockwise around the triangle of pegs in a systematic way.

In 1859 the Irish mathematician Sir William Rowan Hamilton (1805-1865) posed a related problem concerning connected graphs that still remains unsolved. Let's say that a graph has a Hamiltonian path if it has a path that passes through each vertex exactly once (but does not necessarily traverse each arc in the graph). The graph for the Königsberg bridges, for example, has a Hamiltonian path.

Another way of visualizing a solution is to find a Hamiltonian path for an n-dimensional hypercube–a path along the edges of the cube that visits each corner of the cube once and only once. For example, with $n = 3$, we find a Hamiltonian path for a 3-dimensional cube.[2]

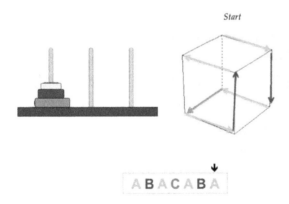

Here let's call the horizonal axis shown by the yellow arrows the x-axis, the vertical axis by the purple arrows the y-axis, and the remaining axis shown by the green arrow the z-axis. The three axes correspond to the colored disks. The Hamiltonian path through the 3-dimensional cube represents a solution to the 3-disk Towers of Hanoi puzzle. The color of each arrow tells you which disk to move and the 3 dimensions—horizontal, vertical, and depth—correspond to the three pegs.

Similarly, a Hamiltonian path through the 4-dimensional cube or tesseract corresponds to the solution to the 4 disk Towers of Hanoi puzzle.

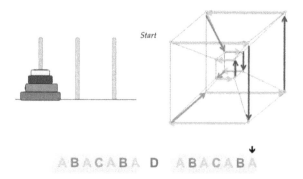

ABACABA D ABACABA

The recursive nature of a solution to the Tower of Hanoi is mirrored in the geometric structure in which there is a small cube within a larger one.

Because of the abstraction involved in modeling, it is easier to see how to generalize our results to other contexts. Euler's theorem, for example, has had applications in fields as diverse as molecular biology and switching circuit analysis, for example. To illustrate the power of abstraction and generalization, let's consider a pair of puzzles.

Below is a map of Euler's pond. There are five islands as well as the land surrounding the point. They are connected by the red bridges. Can you plan a walk–beginning on an island or on the land surrounding the pond that crosses each bridge once and only once? Can you plan such a walk with the extra bridge indicated by the dotted line?

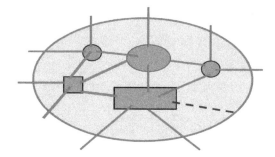

The philosopher Immanuel Kant (1724-1804) lived in Königsberg, Prussia and never left town. He was famous for taking such regular walks at 3:30 in the afternoon that the inhabitants could set their watches by him. Kant is best known for the *Critique of Pure Reason*, which he described as "dry, obscure, contrary to all ordinary ideas, and on top of that prolix."

Below is a floor plan of Kant's home with 16 walls and 5 rooms. In order for Kant's ghost to escape the house, he must walk through each wall once and only once on one continuous path. Can you discover a path that will exorcise Kant's ghost? (Of course, the ghost can't cheat by going over or under a wall or through a corner.)

Can the ghost escape if the top middle wall is knocked down and replaced by the two walls indicated by the broken lines?

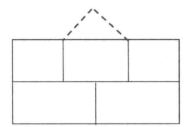

Kant's Ghost Puzzle

Try labeling each of the land areas in Euler's pond puzzle with its degree (not forgetting the surrounding land). Now label each of the rooms in Kant's Ghost puzzle with the number of walls it has. (Consider the outside world a room as well having 9 walls that separate it from Kant's house!). What do you notice about the two numbered diagrams?

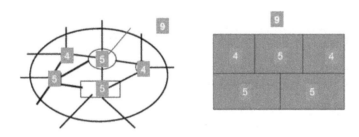

It turns out that the two puzzles have exactly the same structure (in fact, the two structures are, in the mathematician's lingo, isomorphic). Having an abstract model helps us to see more clearly when two real world situations can be solved using essentially the same methods.

Elaborating a Solution

Euler's solution to the famous puzzle of the Königsberg bridges illustrates the elements of problem solving. First, Euler was able to *perceive* an important and general problem in the specific puzzle of the Königsberg bridges. Secondly, Euler, disregarding the *irrelevant* details and isolating the *relevant* elements of the given situation, was able to represent these elements.

Thirdly, Euler critically *examined* the operations with their constraints and considered various problem-solving languages with which to solve the problem.

What Euler did, in effect, was to construct by a *model* or abstract representation of the given situation. Then instead of merely enumerating all the possible solutions paths, Euler was able to use the model to prove a theorem about the necessary and sufficient conditions for the existence of an Euler path.

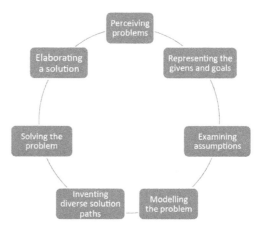

Remember the stages of problem solving using the acronym PREMISE

Euler actually proved something stronger than the result we proved in this chapter. We showed that a connected graph having an Euler path must meet a certain *necessary* condition: *if a connected network has an Euler path, then it must have either zero or exactly two vertices of odd degree*. Euler *also* proved that these conditions were *sufficient* for the existence of the Euler path. So the converse of this conditional is also true: *if any connected network has either none or exactly two vertices of odd degree, then it will have an Euler path*. Can you prove the converse? If you can't there's no better place to go than to Euler's original paper.

Euler not only decisively solved the puzzle of the Königsberg bridges in full generality, but he elaborated his solution in such a way that others could easily follow the steps he took. Euler was known as a master in the art of explanation. His deep concern for explanatory elegance was admired by generations of mathematicians who have followed his lead. The following biographical sketch, which appears on the IBM poster "Modern Men of Mathematics", may give you some idea of the man behind the mathematics:

> LEONHARD EULER (1707-1783) Pious but not dogmatic, Euler conducted prayers for his large household, and created mathematics with a baby on his lap, children playing all around. Euler withheld his work on the calculus of variations so young Lagrange could publish first, and showed similar generosity on many other occasions. Utterly free of false pride, he always explained how he was led to his results: "the path I followed will be of some help, perhaps." Generations of mathematicians followed Laplace's advice: "Read Euler, he is our master in all." Dictating or writing on his slate, Euler kept up his unparalleled output, though totally blind for the last 17 years of his life. He promised to supply the Petersburg Academy with papers until 20 years after his death... The most prolific mathematician in history died while playing with his grandchildren and drinking tea.[3]

Summary of Concepts

A MODEL is an abstract representation of a given situation.

A CONNECTED GRAPH is a model to which Euler's theorem applies.

An EULER PATH is a solution to a connected graph problem.

MODELING helps to make our assumptions explicit, to make our searches more efficient by allowing parallel as opposed to serial searches, and to understand how results might be generalized to other contexts.

The PROBLEM SOLVING PROCESS involves the following kinds of steps or stages:

>**P** erceiving problems–*can this situation be occasion for creativity?*
>**R** epresenting the givens and goals–*what's the real problem?*
>**E** xamining assumptions–*am I assuming something that can be questioned?*
>**M** odeling–*is there a better way to represent the problem, problem space, or operations?*
>**I** nventing–*can I come up with criteria for higher quality solutions?*
>**S** olving the problem–*is there a shorter or more systematic way to solve the problem?*
>**E** laborating–*is my explanation correct, clear, elegant?*

Exercises

Group I

1. Euler's Theorem

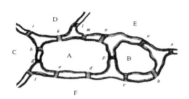

Here is another configuration of bridges and rivers. Abstract a connected graph from the map, and apply Euler's theorem to it to determine whether there is an Euler path.

2. Königsberg Bridge-Crossing Rally

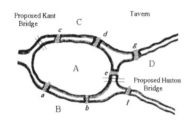

To celebrate the 100th anniversary of Leonhard Euler's birth, the townspeople of Königsberg devised a bridge crossing rally. Here is a map of the seven Königsberg bridges together with two proposed bridges. One of the proposed bridges was named after Kant, who was born in Königsberg in 1794; the other was named after Euler's grandson Huston.

To win the rally a contestant must plan a walk so as to cross each of the bridges exactly once. If you must end your walk at the tavern, which of the following construction proposals will enable you to win–

(A) building only the Kant bridge;
(B) building only the Huston bridge;
(C) building both bridges;
(D) building either the Kant bridge or the Huston bridge but not both;
(E) building neither.

Justify your answer by applying Euler's theorem to the graph for each proposal.

Group II

3. Exit Strategy

According to Egyptian myths, the ghost of King Tut's nephew haunts the maze of the pyramids. A map of the walls of King Tut's pyramids appears to the right. Every night at midnight the ghost of King Tut's nephew traverses the maze. Only after the nephew is able to solve the maze can he rest in peace.

The maze is solved by planning a path that passes through each of the walls of the maze once and only once. The ghost may not go over or under a wall nor can he pass through a corner. In other words, the ghost must plan a path which passes through each wall of the pyramid once and only once. Here are some of the ghost's unsuccessful attempts.

You may have correctly surmised that Euler's theorem can be applied to the maze, but how? The ghost doesn't traverse the maze, but instead he must break through each wall of the maze. Imagine placing a small bridge breaking through each wall. Then you get the corresponding connected graph by reducing the rooms and the outside to vertices and by connecting them with arcs running over each of the bridges.

Traversing this connected graph with an Euler path is the same problem as breaking through the walls of the original maze problem. Solve the problem by applying Euler's theorem to the connected graph. Can the ghost of the nephew of King Tut solve the maze?

The puzzle continues. When you unearth the maze, two of your assistants accidentally destroy two of the walls of the maze, leaving the configuration on the right.

What is the connection between this maze puzzle and the Königsberg bridge puzzle?

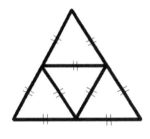

4. *Social Networks*

Dr. Puth, an anthropologist, and his son Eddie, are studying exotic Polynesian islands. For each tribal community he visits, Dr. Puth instructs his son Eddie to constructs a social network. For each of the peaceful communities, Eddie discovers that the number of persons in the social network who have an odd number of relationships is always even. Eddie's father steals Eddie's research and publishes it himself. Eddie is angry and skypes with his mother, a psychoanalyst, who advises Eddie to come home and live with her. She tells Eddie that his father Dr. Puth has written a letter saying that he is leaving the family for a young researcher. Here is Puth's Law:

> A social network is *harmonious* if the number of persons who have an *odd* number of relationships is always *even*.

Show that Puth's Law is not significant empirical discovery.

What is the parity of the sum of all the degrees of vertices in a connected graph? What is Eddie's dilemma?

Group III

5. *Utilities*

The English puzzlist Henry Ernest Dudeney invented one of the oldest puzzles concerning planar graphs, a set of vertices connected by arcs that do not intersect. Each house must receive gas, water, and electricity.

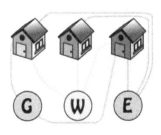

(A) Can lines be drawn to connect each house with each utility in such a way that no line intersects another?

(B) Show that the utilities problem can be solved if it is drawn on a surface that is, not a plane, but a torus (i.e., a donut, or, a coffee cup, which are topologically speaking, identical).

6. *Circuit Boards*

Devising planar graphs is also of technological interest. Printed circuits, for example, will short circuit if any two paths cross.

(A) Create a printed circuit within the rectangle that connects the paired vertices without crossing over one another.

(B) Create a printed circuit on the grid lines that connects paired vertices without crossing over one another.

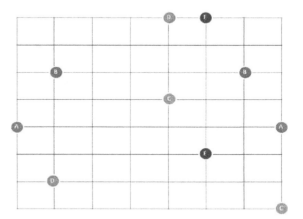

7. *Kuratowski's Theorem*

A graph is planar when it can be drawn on a plane without any of its edges crossing each other. The utilities problem above establishes that the K33 graph—or a graph connecting three vertices to three other vertices—is not planar.

Another well known non-planar graph is the pentagon with a five-pointed star inside, which is known as K_5.

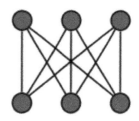

Kuratowski's Theorem [1930] gives the following elegant result:

> A finite graph G can be drawn on a plane without any of its edges crossing each other *if and only if* it not possible to subdivide the edges of K_5 or K_{33} (the utilities graph) and possibly add additional edges and vertices to build a graph isomorphic to G.

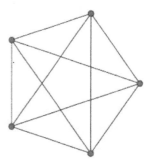

Look up various proofs of Kuratowski's theorem and try to write up your own correct, intuitively convincing, and elegant proof of this theorem.

8. *Hamiltonian Paths*

Today there is no known fast and completely general method for deciding whether a graph contains a Hamiltonian path. However, Hamilton did show that a Hamiltonian path exists for the edges of each of the five regular Platonic solids. Can you find Hamiltonian paths for each of them?

9. *Towers of Hanoi and the Hypercube*

Show how solving the 4 disk Tower of Hanoi problem is related to finding a Hamiltonian path for the tesseract or 4-dimensional hypercube.

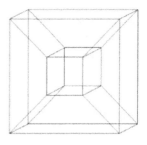

10. *End of the Universe*

Create a spreadsheet to figure out how long it would take to solve a Towers of Hanoi puzzle with n disks.

1 minute
1 hour
1 day
1 year
1 century
1 millennium
1 age (1 million years)
1 epoch (10 ages)
1 era (10 epochs)
1 eon (5 eras)

11. *Ramsey's Friends*

The theorem is named after the Cambridge mathematician Frank Plumpton Ramsey (1903-1930) who died tragically at a young age. Ramsey is known for his mathematical acumen–he formulated ("Ramsified") Russell's "Simple Theory of Types," translated Wittgenstein's Tractatus from German into English, and was instrumental in persuading Ludwig Wittgenstein to return to Cambridge after he (W.) had become a (frustrated) primary school teacher in a small town in Lower Austria.[4]

Ramsey's theorem states that at a party of six people there are either three mutual acquaintances or three mutual strangers. Model this theorem using a bi-chromatic coloring a *complete graph* of a hexagon–a *complete* graph is one in which every pair of vertices is connected with a line. The six persons at the party are represented by the six vertices. Connecting a pair of vertices with a red line means the two persons are strangers; connecting them with a blue line means the two persons are friends.

What does Ramsey's theorem state in terms of coloring the complete graph?

You can't color the complete graph for six vertices without _____. (In fact, you can't do it without making two such triangles!)

I almost always felt, with regard to any subject we discussed, that he [Ramsey] understood it much better than I did, and where (as was often the case) he failed to convince me, I generally thought the probability was that he was right and I was wrong, and that my failure to agree with him was due to my lack mental power on my part.[5]
–G. E. MOORE

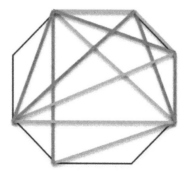

4

Puzzles, Paradoxes & Previews

Method consists entirely in properly ordering and arranging the things to which we should pay attention.
—René Descartes (1596-1650)

In this chapter you'll be challenged with an assortment of puzzles and paradoxes chosen to stimulate your curiosity and to bring into play your sense of invention. Typically, we'll begin with a challenging 'real world' problem, and then follow it with a simpler 'puzzle' problem of the same type. The 'real world' problem will be solved in later modules using the techniques developed more fully there, but the puzzle problems will be solved in this chapter using simplifications of those same methods and models. Puzzles similar to those discussed here are now commonly found on standardized tests. But, besides the utilitarian interest in improving your scores on such tests, these simple problems will provide a preview of coming attractions.

Matrix Logic

To illustrate the usefulness of having a systematic way of representing inference, consider the following matrix logic puzzle:

Who's the Logician?

Mr. Kalish, Mr. Kay, and Mr. Krom are the lawyer, the linguist, and the logician at a conference (but not necessarily in that order). Each of them has one and only one of the occupations listed. Also at the conference are three women who, coincidentally, have the same three last names. (They are identified in what follows by "Ms." prefixed before their last names.) Each of the women lives in only one of the three cities mentioned below, and no two of the women live in the same city. You are also given the following facts:

(A) Ms. Krom lives in Los Angeles.
(B) The linguist lives in New York.
(C) Ms. Kay has never heard of Donald Knuth's classic work in computer science.
(D) The woman who has the same last name as the linguist lives in Paris.
(E) The linguist and one of the women, a computer science professor, live in the same condominium.
(F) Mr. Kalish demolished the lawyer at chess.

Can you deduce who is the logician?

Due to the limits of short term memory, it is difficult to use all the information in the puzzle given above without some notational aid. One such aid is a matrix, a rectangular array of boxes or cells that represents all the relevant possibilities. What are the relevant elements?

Since we have three men, each of whom has one of three occupations, and three women, each of whom lives in exactly one of three cities, it will be convenient to have two 3 x 3 matrices. One matrix will contain entries for the three men and the three occupations; the other matrix will contain entries for the three women and the three cities. Let us see how this works.

The matrices initially look like this:

Mr.	Lawyer	Linguist	Logician
Kalish			
Kay			
Krom			

Ms.	LA	NY	Paris
Kalish			
Kay			
Krom			

We now represent the information contained in the premises by placing 1s and 0s in the cells. Placing a 1 indicates a cell represents that the relevant possibility is actual or true, and placing a 0 in a cell indicates that the relevant possibility is ruled out. Since we are given that each of the men has one and only one of the three occupations, whenever we enter a 1 into the matrix, we can place 0s in the same row and column, and whenever all the other entries in a row or column contains 0s, we may place a 1 in the remaining cell. Since each of the women lives in one and only one of the cities, similar remarks hold for the second matrix.

The first step is to represent those premises that give definite information by themselves. Premise (A), for example, tells us that Ms. Krom lives in Los Angeles. We place a 1 in the cell at the intersection of 'Ms. Krom' and 'Los Angeles'. We annotate this entry with a subscript to indicate the relevant clue or premise from which it was derived. We may also place 0s in the same row and column.

Ms.	LA	NY	Paris
Kalish	0		
Kay	0		
Krom	1_A	0	0

Premise (F), which states that Mr. Kalish beat the lawyer at chess, also gives definite information. Premise (F) eliminates the possibility that Mr. Kalish is the lawyer. Hence, we may put a 0 in the relevant cell.

Mr.	Lawyer	Linguist	Logician
Kalish	0_F		
Kay			
Krom			

The next step is to combine premises which together yield definite information. The clues talk about a computer science professor. Who is this professor? By premise (E) we know that the computer science professor lives in the same condominium, and hence the same city, as the linguist. Premise (B) tells us that the linguist lives in New York. Therefore, the computer science professor must live in New York. We already know that Ms. Krom lives in Los Angeles and so she can't be the computer science professor. But, according to premise (C), neither can Ms. Kay who has never heard of a classic work in computer science.

By a process of elimination, we deduce that Ms. Kalish is the computer science professor and that she lives in New York. We can therefore place a 1 in the cell at the intersection of "Ms. Kay" and "New York", and we may also put 0s in the same row and column.

Ms.	LA	NY	Paris
Kalish	0	1_{EBC}	0
Kay	0	0	
Krom	1_A	0	0

Since Ms. Kay must live in one of the three cities, we may now deduce that Ms. Kay lives in Paris.

Ms.	LA	NY	Paris
Kalish	0	1_{EBC}	0
Kay	0	0	1
Krom	1_A	0	0

We can now make use of premise (D), which relates the information in the two matrices. Premise (D) states that the woman who has the same name as the linguist lives in Paris. We know that Ms. Kay lives in Paris. Therefore, using clue (D) backwards, we may deduce that Mr. Kay is the linguist.

Mr.	Lawyer	Linguist	Logician
Kalish	0_F		
Kay		1_D	
Krom			

Placing 0s in the same row and column as this last entry, the remaining deductions are simple.

Mr.	Lawyer	Linguist	Logician
Kalish	0_F	0	
Kay	0	1_D	0
Krom		0	

Since neither Mr. Kalish nor Mr. Kay is the lawyer, Mr. Krom must be the lawyer. Hence, we may deduce that Mr. Kalish is the logician.

Mr.	Lawyer	Linguist	Logician
Kalish	0_F	0	1
Kay	0	1_D	0
Krom	1	0	0

Matrix logic represents a systematic way of keeping track of deduction by elimination.

Liars, Truth-Tellers & Truth Tables

Logic, like the microscope, takes certain features that are of negligible importance in everyday usage and makes them prominent. For this reason, logic is crucial when precision and accuracy of expression are paramount.

Consider the following transcript from the Watergate Impeachment proceedings against President Richard Nixon.

THE HOUSE JUDICIARY COMMITTEE DEBATE ON "UNLESS" Debate on the Articles of Impeachment (pp. 137-49)

Friday, July 26, 1974
House of Representatives, Committee on the Judiciary.
The committee met, pursuant to notice, at 11:55 a.m.
Sen. Peter W. Rodino, Jr. (Chairman) presiding.

MR. MCCLORY: I have a motion at the clerk's desk, which I have distributed among the members, Mr. Chairman.

THE CHAIRMAN: The clerk will read the motion.

THE CLERK (reading): Mr. McClory moves to postpone for 10 days further consideration of whether sufficient grounds exist for the House of Representatives to exercise its constitutional power of impeachment unless by 12 noon, eastern daylight time, on Saturday, July 27, 1974, the President fails to give his unequivocal assurance to produce forthwith all taped conversations subpoenaed by the committee which

DICTIONARY:
P: The committee should postpone for 10 days further consideration of whether sufficient grounds exist for impeachment proceedings.
Q: By 12 noon on Saturday, July 27, the President gives his unequivocal assurance to produce all taped conversations subpoenaed by the committee.

MCCLORY'S MOTION:
P unless Q fails.

LATTA'S PARAPHRASE:
If Q fails, then P; and if Q, then not P.

LATTA'S CORRECTION:
P provided that Q.

MCCLORY'S IMPLICATION:
'P unless Q fails' implies 'not P if Q fails'

DENNIS'S EQUIVALENCE:
'P unless Q fails' is equivalent to 'P provided that not Q'

The New York Times
August 9, 1974

are to be made available to the district court pursuant to court order in UNITED STATES VS. MITCHELL...

MR. LATTA: ...I just want to call [Mr. McClory's] attention before we vote, to the wording of his motion. You move to postpone for 10 days unless the President fails to give his assurance to produce the tapes. So, if he fails tomorrow, we get 10 days. If he complies, we do not. The way you have it drafted I would suggest that you correct your motion to say that you get 10 days providing the President gives his unequivocal assurance to produce the tapes by tomorrow noon.

MR. MCCLORY: I think the motion is correctly worded, it has been thoughtfully drafted.

MR. LATTA: I would suggest you rethink it...

MR. MANN: Mr. Chairman, I think it is important that the committee vote on a resolution that properly expresses the intent of the gentleman from Illinois [Mr. McClory] and if he will examine his motion he will find that the words 'fails to' need to be stricken and...

MR. MCCLORY: If the gentleman will yield, the motion is correctly worded. It provides for a postponement for 10 days unless the President fails tomorrow to give his assurance, so there is no postponement for 10 days if the President fails to give the assurance, just 1 day. I think it is correctly drafted. There is a 10-day postponement unless the President fails to give assurance. If he fails to give it, there is only a 24-hour or there is only a 23 1/2-hour day [sic].

MR. RANGEL: Mr. Chairman?

MR. MCCLORY: I think the Members understand what they are voting on.

MR. DENNIS: Will the gentleman yield to me?

MR. RANGEL: Mr. Chairman –

MR. DENNIS: The gentleman yielded to me, Mr. Rangel. Excuse me. I know you did not realize that fact.

MR. RANGEL: No; I did not.

MR. DENNIS: He did not. I realize that. What Mr. Mann says and what Mr. Latta says is true, in my opinion. It would be much better if you said 'provided that' or 'unless he does not', or something, but I think nevertheless, the gentleman from Illinois is correct, that although this is a very backhanded way of stating it, it does in fact state it because it says he gets 10 days if he does not–well, it is a backhanded way of stating what the gentleman is trying to state. It could be improved but what he is doing is nevertheless there.

MR. MANN: I guess we can settle for it so long as we all understand it, Mr. Chairman.

THE CHAIRMAN: Will the gentleman yield?

MR. RANGEL: Mr. Chairman, I think this motion itself has provided sufficient delay and I move the question.

THE CHAIRMAN: The question is on the motion of the gentleman from Illinois...

THE CLERK: Mr. Chairman, 11 members have voted aye, 27 members have voted no.

THE CHAIRMAN: And the motion is not agreed to...

Clearly this lengthy debate about the logic of the word 'unless' can be confusing. Yet with the simple technique of truth table, we can begin to find a definitive way of dealing with questions of equivalence or non-equivalent of the various proposals.

The film *Chariots of Fire* (awarded Best Picture in 1981) is based on a true story of two great runners who competed in the 1924 Olympic Games for Great Britain, both winning gold medals. Harold Abrahams (Ben Cross) is the runner who runs on nerves and uses his strength to run his competitors "off their feet" to fight the demons of anti-Semitism he faces during his schooling in Cambridge. Eric Liddell (Ian Charleson) is the "Flying Scotsman" who was born in China to Scottish missionaries runs to express his spiritual devotion. Harold is crushed after a race because he has never lost before, and Sybil, his girlfriend, tries to encourage him.

ERIC LIDDELL
1924 Olympic Games

Abrahams says, "I don't run to take a beating. I run to win. If I can't win, I won't run."
Sybil retorts: "If you don't run, you can't win!"

Modifying this exchange a bit, let's consider two intuitively equivalent sentences:

(1) Harold ***can't*** win ***unless*** he runs.
(2) ***If*** Harold ***doesn't*** run, then Harold ***can't*** win.

Assuming the equivalence of these two statements, we can use truth tables to discover the logic of the word "unless."

To introduce the use of truth tables consider a popular type of logic puzzle that was turned into an art form by the logician Raymond Smullyan in a series of sophisticated and elegant puzzle books.

Suppose you are vacationing on the Island of Crete, which is inhabited by the proverbial two groups–the members of one group (the Knights) always tell the truth, and the members of the other group (the Knaves) always tell falsehoods. You meet two of these inhabitants—Post and Quine—each of whom is a Knight or a Knave.

A *truth table* is a model for evaluating the truth-value of a complex statement on the basis of the truth-values of its simpler parts. The basic idea of using a truth table to solve the above problem was expressed by Sherlock Holmes:

"How often have I told you that whenever you have eliminated the impossible, whatever remains - however improbable - must be the truth!"
 A Study in Scarlet
—ARTHUR CONAN DOYLE (1859-1930)

Post, referring to both himself and Quine, says, "Neither of us is a Knight." Can you deduce to which group each of them belongs?

Try to figure the answer out for yourself. If you already have the solution, congratulations! However, we are not interested in the solution itself but in a systematic method of arriving at the solution no matter how complex.

A fundamental property of such complex sentences is that their truth-values are completely determined by the truth-values of their component simple sentences. This property is known as *truth-functionality*.

Given a sentence φ, another sentence called the negation of φ can be formed by prefacing the sentence with the words 'it is not the case that.' Symbolically, the negation of a sentence φ is written '∼φ'. If φ is true, then ∼φ is false; and if φ is false, then ∼φ is true. In other words, the truth-value of a negation is opposite to the truth-value of the negated sentence.

Any pair of sentences can be combined with the word 'and' to form a complex sentence called a *conjunction*. Symbolically, the conjunction of a sentence φ and a sentence ψ is written '(φ ∧ ψ)', and φ and ψ are called *conjuncts*. A conjunction is true if both its conjuncts are true; otherwise, it is false.

Similarly, any pair of sentences can be combined with the word 'or' to form a complex sentence called a *disjunction*. Logicians assume "or" to be inclusive, unless otherwise stated. Symbolically, the disjunction of a sentence φ and a sentence ψ is written '(φ ∨ ψ)', and φ and ψ are called *disjuncts*. A disjunction is true if one or the other or both disjuncts are true. Alternatively, the disjunction is false if both its disjuncts are false; otherwise, it is true. The *exclusive* 'or', which in Latin is the word '*aut*' is true if either one or the other, but not both disjuncts are true. The exclusive 'or' can be expressed in terms of the foregoing logical symbols:

$$(P \vee Q) \wedge \sim(P \wedge Q)$$

The expression 'neither P nor Q' can be expressed using the above logical systems:

$$\sim(P \vee Q)$$

or, alternatively,

$$\sim P \wedge \sim Q .$$

The symbol '∨' comes the Latin word '*vel*', which is 'or' in the *inclusive* sense.

Any pair of sentences can be combined with the phrase 'if...(then)...' to form a complex sentence called a *conditional*. The sentence introduced by the word 'if' is called the *antecedent* and the sentence introduced by the word 'then' is called the *consequent* of the conditional. Symbolically, a conditional formed with a sentence φ as antecedent and a sentence ψ as consequent is written '(φ → ψ)'. The conditional (φ → ψ) is false if the antecedent φ is true and the consequent ψ is false; otherwise the conditional is true. Notice that this means that the conditional is true if the antecedent is false or if the consequent of true.

Finally, any pair of sentences can be combined with the phrase 'if and only if' to form a complex sentence called a *biconditional*. The components of a biconditional are called *constituents*. Symbolically, a biconditional formed from sentences φ and ψ is written '(φ ↔ ψ)'. The biconditional (φ ↔ ψ) is true if both constituents have the same truth-value; otherwise it is false.

The phrase 'if and only if' is sometimes abbreviated '*iff*.'

The above truth rules can be summarized in a *truth table*. The four rows in the first two columns list all the possible combinations to the sentence letters 'P' and 'Q', and the remaining columns list the truth-values for the various complex sentences in terms of those initial four assignments:

P	Q	~P	P ∧ Q	P ∨ Q	P → Q	P ↔ Q
T	T	F	T	T	T	T
T	F	F	F	T	F	F
F	T	T	F	T	T	F
F	F	T	F	F	T	T

Truth Table Definitions for the Logical Connectives

To learn the truth tables forwards and backwards, you don't need to memorize 20 individual values. Focus on the places where the values change. Here are the essentials to commit to memory:

$$\sim \quad \text{opposite}$$
$$\wedge \quad \text{T iff T} \wedge \text{T}$$
$$\vee \quad \text{F iff F} \vee \text{F}$$
$$\rightarrow \quad \text{F iff T} \rightarrow \text{F}$$
$$\leftrightarrow \quad \text{T iff same}$$

(The abbreviation '*iff*' for '*if and only if*' by '*iff*' is due to Paul Halmos.)

With these simple tools in hand, we can actually compute an answer to our original Knight and Knave puzzle. Let's introduction a scheme of abbreviation or dictionary:

78   THINKING MATTERS: CRITICAL THINKING AS CREATIVE PROBLEM SOLVING

P: Post is a Knight
Q: Quine is a Knave

Now, in the above puzzle, Post says, "Neither of us is a Knight." We can symbolize what Post said as follows:

$$\sim(P \vee Q)$$

We are not told whether Post is a Knight or a Knave, but we are told that Post is a Knight if and only if what Post said is true. In other words, we know that the following complex sentence is true:

$$P \leftrightarrow \sim(P \vee Q)$$

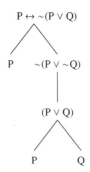

A *grammatical tree*, which is a systematic way of listing all the simpler components of the complex sentence, can be useful guide to the order of steps in filling out a truth table.

The above complex sentence is a biconditional with constituents 'P' and '$\sim(P \vee Q)$'. This latter constituent is a negation of the disjunction of P and Q. We can represent the complex structure of this sentence in a grammatical tree.

As a guide to filling out a truth table, we simply number the nodes from bottom to top using the following rules:

* Begin the numbering by assigning the same number to all occurrences of the sentence letters at the bottom of the tree.
* Then you can number a node provided that all the nodes below it are already numbered.

The first step is to copy the values assigned to the sentence letter P in the left-hand column into all the columns beneath the occurrences of P in the sentence for which we are constructing the truth table (label them all '1' in the row for the *Steps*).

P	Q	[P	↔	~	(P	∨	Q)]
T	T	T			T		
T	F	T			T		
F	T	F			F		
F	F	F			F		
Steps		1			1		

The second step is to do the same for the occurrence of Q (label this step '2' in the row for the *Steps*).

P	Q	[P	↔	~	(P	∨	Q)]
T	T	T			T		T
T	F	T			T		F
F	T	F			F		T
F	F	F			F		F
Steps		1			1		2

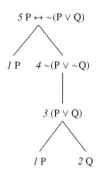

The *numbering* of the nodes of the grammatical tree is a guide to the order of filling out the columns of the truth table.

Now we are ready to climb the tree one step at a time using the truth rules for the logical operators and connectives. We can fill in the column for an operator or connective only if all the columns for the nodes below that operator or connective have been filled.

The values for step 3 comes from comparing columns 1 and 2 the rule for ∨ . The rule for the disjunction is that ∨ is F if and only if both disjuncts are F. So in the column under the '∨ ', we place an F in the fourth row of that column and T in the other three rows.

P	Q	[P	↔	~	(P	∨	Q)]
T	T	T			T	T	T
T	F	T			T	T	F
F	T	F			F	T	T
F	F	F			F	F	F
Steps		1			1	3	2

The next step is to fill in the values for the negation ~ (column 4) using the rule for negation. The rule for negation ~ is that it gives the *opposite* value of the sentence it negates. The values of column 4 reverse, not the values in column 1, but rather of the values in column 3. Column 4 is the negation of the disjunction (column 3) not the negation of sentence letter P (column 1).

P	Q	[P	↔	~	(P	∨	Q)]
T	T	T		F	T	T	T
T	F	T		F	T	T	F
F	T	F		F	F	T	T
F	F	F		T	F	F	F
Steps		1		4	1	3	2

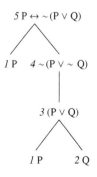

Numbered Nodes

The last step is to fill in the truth values for the biconditional (column 5). The biconditional ↔ has the value T if and only if both constituents have the *same* truth value. The constituents of the biconditional are given in columns 1 (the sentence letter P) and 4 (the negation of the disjunction). These two have the same truth value (the value F) in the third row of the truth assignments.

P	Q	[P	↔	~	(P	∨	Q)]
T	T	T	F	F	T	T	T
T	F	T	F	F	T	T	F
F	T	F	T	F	F	T	T
F	F	F	F	T	F	F	F
Steps		1	5	4	1	3	2

How does this truth table compute the answer to our original puzzle?

The puzzle conditions are met when Post is a Knight if and only what he said is T. We constructed a truth table for this the biconditional and found it is T in just one case, namely, in the third row of the truth assignments where P has the value F and Q has the value T.

P	Q	[P	↔	~	(P	∨	Q)]
T	T	T	F	F	T	T	T
T	F	T	F	F	T	T	F
F	T	F	T	F	F	T	T
F	F	F	F	T	F	F	F
Steps		1	5	4	1	3	2

Since 'P' stands for the statement "Post is a Knight" and 'Q' for the statement "Quine is a Knight", the puzzle is solved when P has the truth value F and Q has the truth value T, i.e., when it is false that Post is a Knight and it is true that Quine is a Knight.

The solution is that Post is a Knave and Quine is a Knight.

You probably were able to figure out this puzzle in your head. So why bother with the truth table? The reason for doing so is that we now have a systematic way of solving all such problems. We can even program a computer to solve the problem for us!

For example, the following puzzle was posed by Eben Kadile and solved using a computer program to do "brute force" work:

(1) Alfred says, "Boole is a Knave if Church is a Knight";
(2) Boole says, "I am a Knight if and only if Alfred is a Knight";
(3) Church says, "Dummett is a Knave or Alfred is a Knave";
(4) Dummett says, "Alfred is a Knave";
(5) Egbertus says, "Dummett is a Knight and I am a Knight";

(6) Frege says "I am a Knight";
(7) Gödel says "Frege is a Knave";
(8) Hilbert says "I am a Knight if either Gödel is a Knight or Alfred is a Knight";
(9) Jan says, "Frege is a Knave if Boole is a Knave";
(10) Kripke says, "Jan is a Knave or Egbertus is a Knight".

Writing out the complete truth table for n speakers requires 2^n rows! So the manual calculation of a truth table can become tiresome quickly. When we learn the art of deductive logic in Module II, we'll see how to logically *deduce*, rather than tediously *calculate*, the solution to such puzzles.

Let's consider an example with three individuals–Art, Gwen, and Lance are each of whom is either a Knight or a Knave.

Gwen says, "Art and Lance are different types."
Art says, "Gwen is a Knave"

Can you determine whether Lance is a Knight or a Knave?
Now what Gwen means by saying Art and Lance are *different* types is that it's not the case that Art are Lance are the *same* type, i.e., it is not the case that (Art is a Knight if and only if Lance is a Knight).

This puzzle is interesting for several reasons: (1) the problem can be solved using an 8 row truth table; (2) there are natural and logically equivalent ways of expressing what Gwen said; (3) there isn't information to determine what Art and Gwen are, but we can definitely determine what Lance is!

First we introduce a scheme of abbreviation or dictionary:

A: Art is a Knight
G: Gwen is a Knight
L: Lance is a Knight.

Second, we symbolize what Gwen and Art said, and form the biconditionals expressing what we know, namely, that a speaker is a Knight if and only if what they said is true:

G ↔ ~(A ↔ L)
A ↔ ~G.

Here the 8 row truth table is intentionally left blank so you can calculate the solution for yourselves. A grammatical tree is provided in the margin as a guide to filling out the columns. The final column has two entries with the value T.

82 THINKING MATTERS: CRITICAL THINKING AS CREATIVE PROBLEM SOLVING

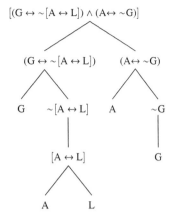

A	G	L	[(G	↔	~	[A	↔	L])	∧	(A	↔	~	G)]
T	T	T							F				
T	T	F							F				
T	F	T							T				
T	F	F							F				
F	T	T							T				
F	T	F							F				
F	F	T							F				
F	F	F							F				
Steps									8				

However, in both of the rows where this happens, we have the proposition L that "Lance is a Knight" has the value T. Hence, in either case, Lance is a Knight. It turns out that we can't deduce whether Art and Gwen are each individually a Knight or Knave, but we can deduce that exactly one of them is a Knight and the other a Knave. Art and Gwen are different types.

Use truth tables to verify the following are logically equivalent:

(1) ~(P ↔ Q)
(2) (P ↔ ~Q)
(3) (~P ↔ Q)
(4) (P ∨ Q) ∧ ~(P ∧ Q).

The last formula symbolizes the *exclusive 'or'* in English which is paraphrased by "Either P or Q, but not both P and Q." Here again rather than carrying out the tedious calculations, we could have deduced solution by knowing a few logical theorems and laws of inference.

"Unless" Revisited

This method of truth table analysis can be used to analyze the House Judiciary Committee's debate on the logic of 'unless' as well as other intriguing questions about the logic of English. Earlier we noted that the following two sentences are intuitively equivalent:

(1) Harold can't win unless he runs.
(2) If Harold doesn't run, then Harold can't win.

Working backwards and forwards, we construct a truth table to discover the logic of *"unless"*. Step 3 obtains by the rule for the conditional. Since we are assuming the logical equivalence of the two statements, place 'T's in the column for step 6. Working backwards, we can deduce that column 5's values must be the same as those of column 3. Step 4 for the negation of W reverses the values of step 1.

W	R	(~	W	unless	R)	↔	(W	→	R)
T	T	F	T	T	T	T	T	T	T
T	F	F	T	F	F	T	T	F	F
F	T	T	F	T	T	T	F	T	T
F	F	T	F	T	F	T	F	T	F
Steps		4	1	5	2	6	1	3	2

A Truth Table to Compute 'Unless'

Now we ask, which logical connective for "unless" has the values of step 5 (in blue) given the values in steps 4 and 2 (in red)?

In the discipline of linguistics the literature of "unless" is quite extensive and interesting. Some logicians treat *unless* as a stylistic variant of *or*, and others as a variant of *if not*, and still others, of *if and only if*, when the disjunction is exclusive, rather than inclusive.[1]

Inference Rules & Fallacies

Lewis Carroll is the pen name of the Oxford mathematician Charles Dodgson, the author of *Alice's Adventures in Wonderland*, and *Through the Looking Glass*. Philosophers and logicians have been intrigued by the rich skein of logic that is woven in these 'children's tales'. Carroll's nonsense is not nearly as random and pointless as it might seem to a child. Much of the 'nonsense' is really a parody of sense as seen through the looking class of symbolic logic.

Alice is playing croquet with the Red Queen, who is notorious for her impatience, and her constant command "Off with his head!" After Alice's encounter with the Cheshire Cat in the forest, the Cat comes back to haunt her while she is playing croquet in the form of a looming disembodied head. The Cheshire Cat, of course, begins to annoy the Red Queen, who order, "Off with his head!" This poses a logical dilemma for Alice.

> When she got back to the Cheshire Cat, she was surprised to find quite a large crowd collected round it: there was a dispute going on between the executioner, the King, and the Queen, who were all talking at once, while all the rest were quite silent, and looked very uncomfortable.
>
> The moment Alice appeared, she was appealed to by all three to settle the question, and they repeated their arguments to her, though, as they all spoke at once, she found it very hard indeed to make out exactly what they said.

The executioner's argument was, that you couldn't cut off a head unless there was a body to cut it off from: that he had never had to do such a thing before, and he wasn't going to begin at *HIS* time of life.

The King's argument was, that anything that had a head could be beheaded, and that you weren't to talk nonsense.

The Queen's argument was, that if something wasn't done about it in less than no time she'd have everybody executed, all round. (It was this last remark that had made the whole party look so grave and anxious.)

It turns out that the arguments in this passage illustrate some common patterns of argument:

Executioner's Argument (*modus tollens*) The Cheshire Cat's head can be beheaded only if it is attached to a body. The Cheshire Cat's head is not attached to a body. Therefore, it cannot be beheaded.

King's Argument (*modus ponens*) If the Cheshire Cat has a head, then it can be beheaded. The Cheshire Cat has a head. Therefore, it can be beheaded.

Queen's Argument (*argument ad baculum*): If something isn't done about it beheading the Cheshire Cat in less than no time, then I'll have everybody executed!

The King's argument has the form known as *modus ponens*. It has the same form as the following well-worn argument:

> If Socrates is a man, then Socrates is mortal.
> Socrates is a man.
> Therefore, Socrates is mortal.

Modus ponens means "the method of placing".

What's important in deductive logic is not the *content* of the argument, but its *logical form*, which can be abstracted as follows:

$$\frac{\varphi \to \psi}{\psi}$$

Modus ponens is a valid argument form and so any instances of the form—whether it is the hackneyed example about Socrates's mortality or the King's logic-chopping argument about beheading the Cheshire Cat—is thereby validated by virtue of its logical form.

The Executioner's argument has the form known as *modus tollens*. It has the same form as the following argument:

Modus tollens means "the method of taking away".

> If Socrates is a sophist, then Socrates teaches for remuneration.
> Socrates doesn't teach for remuneration.
> Therefore, Socrates isn't a sophist.

The valid logical form common to both this argument and the executioner's can be set forth abstractly as follows:

$$\varphi \to \psi$$
$$\underline{\sim \psi}$$
$$\sim \varphi$$

Any argument of this form is also valid–including the one attributed to President Harry S. Truman:

> If Nixon's honest, I'm a monkey's uncle.
> Therefore, Nixon's not honest.

Whereas *modus ponens* and *modus tollens* are valid argument forms, there are counterfeit, or *fallacious* conditional arguments, with a deceptively similar form. Consider the following fallacious argument:

> "My fellow Americans, if you want to 'cut and run' i.e., abandon our troops, then you should vote for my opponent. But real Americans don't want to 'cut and run'. So I ask for your vote on election day!"

These arguments have roughly the same form–give or take a few hidden assumptions.

> If you want to 'cut and run', then you should vote for my opponent.
> You don't want to 'cut and run' (as a true American).
> Therefore, you shouldn't vote for my opponent (i.e., vote for me!)

Here is an obviously fallacious argument with the same logical form:

> If Lincoln died of old age, then Lincoln is dead   (True)
> Lincoln didn't die of old age.   (True)
> Therefore, Lincoln isn't dead!   (False)

Intuitively, this latter argument is invalid because dying of old age isn't the only way of ending up dead. Lincoln was, after all, assassinated.

An argument is deductively valid by virtue of its form. Moreover, an argument form is valid if there are no instances in which the premises are true but the conclusion is false. The counterexample about Lincoln, therefore, establishes the logical form shared by the examples is invalid. This invalid logical form is known as the *fallacy of denying the antecedent*. It is deceptively similar to the valid form of argument *modus tollens* except that it reverses the conclusion and second premise. Unless the first two arguments have some validating logical form, the counterexample about Lincoln establishes that they are invalid or fallacious.

A *counterexample* can be used to establish the invalidity of an argument. To do this we find an instance of the logical form known as *fallacy of affirming the consequent* whose premises are clearly true while its conclusion is clearly false. We can adapt the above example, to construct a counterexample to the fallacious form of affirming the consequent:

If Lincoln died of old age, then Lincoln is dead	(True)
Lincoln is dead.	(True)
Therefore, Lincoln died of old age!	(False)

To summarize, valid argument forms validate each of their instances. If an argument has *some* valid argument form, then it is valid. On the other hand, an argument is invalid if it has *no* valid argument form. Showing invalidity is therefore more involved. You can show an argument form to be invalid by giving one counterexample to its form–that is, an instance of the form with true premises and false conclusion. However, to show an argument to be invalid, it is not sufficient to show that it has *some* invalid argument form. You must show that it has no logically valid argument form. In practice, this means you must give an argument to show that you've correctly represented the logical form of the argument before proceeding to your counterexample.

A valid argument that is an instance of a *valid* argument form might also be an instance of a *fallacious* argument form. Consider for example the (annoying) truisms: *"Whatever will be, will be"* (que sera, sera) or *"It is what it is."* These tautological statements have the implicit form of *modus ponens*: from $P \to P$ and P to deduce P, but these arguments, though valid, are not particularly cogent.

Clearly these arguments don't prove much because they assume in their premises what they want to show in their conclusions. The arguments aren't cogent or convincing. However, we can still ask about their deductive validity or invalidity. Notice the arguments are instances of both *modus ponens* and the fallacy of *affirming the consequent*.

A valid argument has *some*, or *at least one*, validating argument form, but an invalid argument has *no* validating argument form. A valid argument form validates *all* of its instances. The tautological arguments above are valid. What this example teaches us is that demonstrating an argument to be invalid critically depends on having correctly represented its logical form.

Consider an example from Catholic theologian Bernard Lonergan.

> The existence of God is known as the conclusion to an argument and, while such arguments are many, all of them, I believe, are included in

the following general form. If the real is completely intelligible, God exists. But the real is completely intelligible. Therefore, God exists.[2]

We may represent Lonergan's argument as follows:

1. If reality is completely intelligible, then God exists.
2. Reality is completely intelligible.
3. Therefore, God exists.

Lonergan's argument is valid. It has the form of *modus ponens*. However, Lonergan's argument can still be challenged.

An argument is *sound* if it is valid and has true premises.

Is it true that reality is "*completely* intelligible?" If quantum mechanics is the ultimate theory of reality, then perhaps reality is not completely intelligible. The argument is not unquestionably sound.

So let's try to qualify, or hedge, the premises of Lonergan's argument:

1A. If reality is intelligible, then God exists.
2. Reality is intelligible.
3. Therefore, God exists.

This modified argument has the form of *modus ponens*, but its second premise is less questionable. Reality is remarkably intelligible, if not completely so.

But now the first premise isn't one that everyone would readily accept: perhaps reality is intelligible, but how does this imply that God exists. There might be other ways of accounting for the fact of intelligibility other than positing the existence of God.

Consider two possibilities for the first premise:

1A. If reality is intelligible, then God exists.
1B. If God exists, then reality is intelligible.

Now the atheist can object that premise 1A is question-begging because Lonergan is assuming what he needs to show. Its converse, premise 1B, is not question-begging. If God is the creator of the universe and God, at least as traditionally conceived, is supremely rational, then it is reasonable to suppose that God would create an intelligible reality.

Replacing the original first premise with this less question-begging converse, the argument becomes:

1B. If God exists, then reality is intelligible.
2. Reality is intelligible.
3. Therefore, God exists.

Now the argument, though less question-begging, has a different problem: it is invalid.

Let's summarize the four conditional argument forms in a chart:

Valid Forms	Fallacious Forms
$\varphi \to \psi$ φ ——— ψ *Modus Ponens*	$\varphi \to \psi$ ψ ——— φ *Fallacy of Affirming the Consequent*
$\varphi \to \psi$ $\sim \psi$ ——— $\sim \varphi$ *Modus Tollens*	$\varphi \to \psi$ $\sim \varphi$ ——— $\sim \psi$ *Fallacy of Denying the Antecedent*

As a parting application of these argument forms and fallacies, consider the following problem due to Peter Wason and Philip Johnson-Laird [3]. Suppose you are shown four cards that appear as follows:

You are told that each card has a *letter* on one side and a *number* on the other. You are then given the rule:

If a card has a vowel on one side, it has an even number on the other.

Which cards–and only which ones—must you turn over to prove the rule? You have to choose all at once.

Take some time and record your initial answer. In Wason and Laird's first study only 5 of his 128 subjects got the intended right answer!

Peter Wason Test

Let's adopt the following scheme of abbreviation or dictionary:

V: The card has a vowel on one side.
T: The card has an even number on one side.

Now we can symbolize the hypothesis we're testing as the conditional:

$$V \rightarrow T$$

Let's line up this hypothesis with the empirical information we are given:

A	C	4	7
$V \rightarrow T$	$V \rightarrow T$	$V \rightarrow T$	$V \rightarrow T$
V	$\sim V$	T	$\sim T$
T	$\sim T$	V	$\sim V$

Four Conditional Arguments Forms in the Wason Puzzle

Did you notice that only two of the conditional arguments are valid—the first (*modus ponens*) and the fourth (*modus tollens*)? Most people correctly choose the first card but many neglect to choose the last card. Why? Cognitive scientists explain this error as due to *confirmation bias*, the tendency to always try to *confirm* that we're right rather than checking to see if we're wrong (i.e., subjecting our beliefs to *disconfirmation*). If you choose a card associated with a fallacious inference, then perhaps it is useful to be more aware of the fallacies of affirming the consequent and denying the antecedent.

But the experiment is not over. Let's change the puzzle from this abstract and logical one to a setting that is more realistic and connected to everyday experience. You are told that each card has a person's age on one side and what they are drinking on the other. You are then given the rule:

If a person is drinking Vodka, then they are over 21 years of age.

Which cards—and only which ones —*must* you turn over to prove the rule? You have to choose all at once.

Why is it so much easier to reason in a real world context?

Did you find this puzzle far easier to solve? Most people do. The question is why. Both puzzles have exactly the same logical form. We won't stop now to consider the various intriguing explanations. When you learn the rules of logic, it will be much easier for you to see the logical forms, and the logical similarities, in the abstract structure of these puzzles.

Logical Deductions & Tree Diagrams

Sherlock Holmes is one of the most famous characters of English fiction. In the stories by Arthur Conan Doyle, Holmes is often portrayed as making brilliant 'deductions'. In *A Study in Scarlet*, Sherlock Holmes makes the following startling inference, and then explains his process of reasoning to the amazed Doctor Watson:[4]

"Holmes Pulled Out His Watch"
—Sidney Paget

> "Those rules of deduction laid down in that article which aroused your scorn are invaluable to me in practical work. Observation with me is second nature. You appeared to be surprised when I told you, on our first meeting, that you had come from Afghanistan."

> "You were told, no doubt."

> "Nothing of the sort. I knew you came from Afghanistan. From long habit the train of thoughts ran so swiftly through my mind that I arrived at the conclusion without being conscious of intermediate steps. There were such steps, however. The train of reasoning ran, 'Here is a gentleman of a medical type, but with the air of a military man. Clearly an army doctor, then. He has just come from the tropics, for his face is dark, and that is not the natural tint of his skin, for his wrists are fair. He has undergone hardship and sickness, as his haggard face says clearly. His left arm has been injured. He holds it in a stiff and unnatural manner. Where in the tropics could an English army doctor have seen much hardship and got his arm wounded? Clearly in Afghanistan.' The whole

train of thought did not occupy a second. I then remarked that you came from Afghanistan, and you were astonished."

What kind of reasoning is Holmes using here? In Module Two, we'll distinguish between deductive and inductive reasoning and develop a method for diagramming the dependent relations in such pieces of logical reasoning. We'll also learn the principles for setting forth the arguments contained in such passages in a clear and concise way, the standards by which arguments are evaluated, and how to identify common patterns of correct reasoning and how to detect common errors in reasoning or fallacies.

For the moment, let's see how we might go about representing the logical relations in such pieces of reasoning. Consider the following simple sequencing problem.

Graduation

Your goal is graduation. It turns out that you can graduate from college if you have *both* a major and a minor or if you have lunch with the President. Now if you meet the statistics requirement, you can get a minor. And you can get a major if you satisfy *both* the language and the lab requirements. If you have a parking permit, you can *both* meet the lab requirement *and* have lunch with the President. With a registration pack you can *either* meet the lab requirement *or* get a parking permit (but you can't be sure of which). Given this information, which of the following will *guarantee* that you will graduate from college?

(A) a parking permit
(B) meeting the language and the statistics requirements
(C) meeting the language, lab, and statistics requirements
(D) a registration pack and meeting the statistics requirement
(E) a registration pack and meeting the language and statistics requirements
(F) meeting the language and lab requirements

Stop and take a few minutes to try to solve this problem on your own before reading on.

This *sequencing problem* can be solved using a *directed graph* to represent different kinds of dependent relations. We can define our types of dependent relations (or 'T-bars') using the distinction between "*and*" and "*or*" and the distinction between joins and junctions. Crossing these two distinctions gives us four types of T-bars.

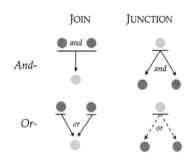

- The "*and*-Join" occurs when two elements join together to yield a third.
- The "*and*-Junction" occurs when one element allows you to obtain two other elements.
- The "*or*-Join" occurs when two elements separately or independently allow you to obtain a third.
- The "*or*-Junction" occurs when one element allows you to obtain either one element or another (possibly, but not necessarily, both).

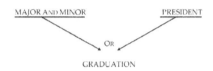

All four types of T-bars occur in the *Graduation* problem.

We're told that you can graduate from college if you have *both* a major and a minor *or* if you have lunch with the President. This information is graphed using an AND-JOIN and an OR-JOIN.

We see below that by grafting the relevant T-bars onto our graph, if you meet the statistics requirement, you can get a minor. And you can get a major if you satisfy *both* the language and lab requirements.

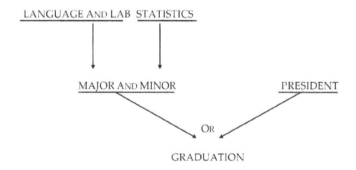

An AND-JOIN will represent the next piece of information, namely that if you have a parking permit, you can *both* meet the lab requirement *and* have lunch with the President:

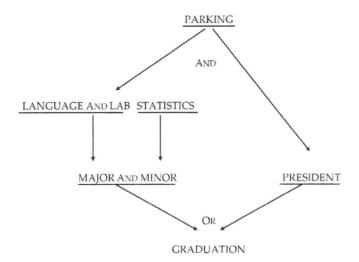

The last piece of information is represented as an OR-JOIN. We're told that with a registration pack you can either meet the lab requirement or get a parking permit (but you can't be sure of which). Putting it all together, and unabbreviating the labels, we finally obtain:

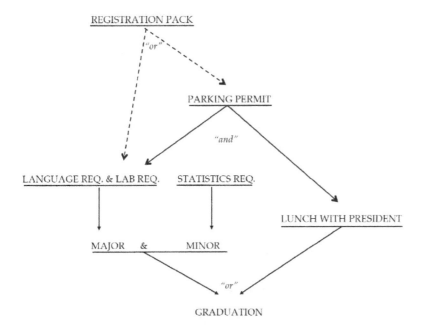

Given all this information, we now wish to know which of the following will guarantee that you will graduate from college. Notice that the directed graph not only represents the given information, but it represents it in a way that gives us insight into the system of dependent relations.

Option (A) Guarantees graduation since a parking permit gets you lunch with the President, which, in turn, ensures graduation.

Option (B) Fails. (Although if you had also had met the lab requirement, you would have gotten the major, and since meeting the statistics requirement gets you the minor, you would have graduated since you would have gotten both the major and the minor.)

Option (C) Guarantees graduation since you would then have both the major and the minor.

Option (D) However, fails to guarantee graduation. If you're lucky enough to get the parking permit, then you'll have lunch with the President, and so you'll be graduated. But perhaps the registration pack will only mean that you satisfy the lab requirement, in which case graduation isn't guaranteed.

Option (E) On the other hand, does guarantee graduation. There are two cases to consider. Having the registration pack guarantees either the parking permit or that you'll satisfy the lab requirement. If you get the parking permit, then, as with option (D) you'll have lunch with the President, and you're home free. On the other hand, suppose the registration pack only means you satisfy the lab requirement. Since under option (E) you've also met the language requirement and the statistic requirement, you'll have both the major and the minor and so will graduate. So either way, you're guaranteed to be graduated.

Option (F) Guarantees you'll have a major, but since you're not guaranteed a minor, graduation will not be guaranteed.

The directed graph not only makes it clear that options (A), (C), and (E) guarantee graduation, but this model represents the information in such a way that it is easy to see what would be required in addition to the failed options to ensure graduation.

We can combine logical inference rules and directed graphs to diagram logical deductions.

- The "*and*-JOIN" combines two propositions to obtain a third.
    - For example, both the premises $(P \rightarrow Q)$ and P are required to obtain Q by *Modus Ponens* [MP]. Similarly, both $(P \rightarrow Q)$ and $\sim Q$ are required to obtain $\sim P$ by *Modus Tollens* [MT].
    - Another example of this kind of relation is known as *Modus Tollendo Ponens* [MTP], which means the "method of taking away in order to place". From the premises $(P \vee Q)$ and $\sim P$, we can obtain Q; and from the premises $(P \vee Q)$ and $\sim Q$ we can obtain P.
    - A third example is to obtain from the two conditionals $(P \rightarrow Q)$

and (Q → P) the biconditional (P ↔ Q) by the inference rule known as *Conditional/Biconditional* [CB].

- The "*and*-JUNCTION" occurs when one proposition allows you to infer separately two other propositions.
  - For example, the premise (P ∧ Q) allows you to obtain both P and Q by a logical inference rule known as *Simplification* [S]. The premise "Roses are red and violets are blue" allows you to deduce "Roses are red" and to deduce "Violets are blue".
  - Another example of this kind of relation is known as *Biconditional/Conditional* [BC]. From the biconditional (P ↔ Q) you may obtain separately the conditional (P → Q) and the conditional (Q → P).
- The "*or*-JOIN" occurs when two propositions separately or independently allow you to infer a third.
  - For example, the premises P and Q each separately allow you to deduce (P ∨ Q). For example, if you have the separate premises "Roses are red" and "Violets or blue" you may infer that "Roses are red or violets are blue" from either of these premises by the logical inference rule know as *Addition* [ADD].
- The "*or*-JUNCTION" occurs when one proposition allows you to infer either one proposition or another (or possibly, but not necessarily, both).
  - For example, the premise (P ∨ Q) allows you to infer either that P is the case or Q is the case (or possibly both). The premise "either Roses are red or violets are blue" allows you to infer that either that roses are red or that violets are blue, but you may not know which of the alternatives is true (or whether both alternatives are true). This kind of relationship occurs when one uses the form of derivation known as *Separation of Cases* [SC] also known as the *Dilemma* argument form.

	JOIN	JUNCTION
And–	MP, MT S MTP CB	ADJ BC
Or–	ADD	SC

Can you model the following deduction by Agatha Christie's Miss Marple? (I am indebted to Andrew Hsu for composing an original version of this problem.)

Arthur would have had no reason to do it unless Aunt Agatha had changed her will before she died. Since she didn't change her will, Arthur didn't do it. That leaves my cousin Dorothy and my baby brother George. Georgie is devious for his age, but he could have murdered Aunt Agatha only if he was strong enough to hold her under water long enough to kill her. Besides, since he isn't even tall enough to reach over the edge of the ornamental fishpond, he couldn't have done it. So, I'm afraid, my dear cousin Dot must have done Aunt Agatha in.

Dictionary:

A. Arthur killed Aunt Agatha
D. Cousin Dorothy ('Dot') killed Aunt Agatha
G. Georgie (could have) killed Aunt Agatha
R. Arthur had a reason to kill Aunt Agatha
W. Agatha changed her will before she died
S. Georgie is strong enough to hold Aunt Agatha under water long enough to kill her
T. Georgie is tall enough to reach over the edge of the ornamental fishpond

Here's a diagram of Miss Marple deduction. (The premises in blue represent implicit premises–premises that are assumed but not stated.) Once you learn the art of deduction, you'll be able to make such deductions in one fell swoop like Miss Marple.

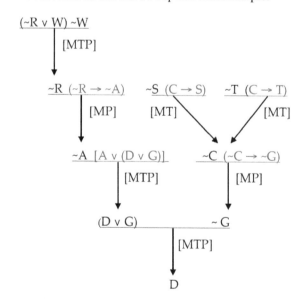

Syllogisms & Venn Diagrams

Traditional Aristotelian logic is limited to simple arguments known as syllogisms (< Gr. συλλογισμός, *syllogismos* meaning "conclusion" or "inference">). A syllogism consists of two premises and a conclusion, each of which is a categorical proposition. There are four types of categorical propositions:

A	universal affirmative	All Sophists are Philosophers.
E	universal negative	No Sophists are Philosophers.
I	existential affirmative	Some Sophist is a Philosopher.
O	existential negative	Some Sophist is not a Philosopher.

The code letters for the four types of categorical propositions come from the first two vowels in AFFIRMO and the vowels in NEGO. In the spirit of playfulness, we may add a fifth vowel:

U *Only* Sophists are Philosophers.
   (or, '*None but* Sophists are Philosophers'.)

The U statement "Only Sophists are Philosophers" is equivalent to "All *non*-Sophists are *non*-Philosophers", which, in turn is equivalent to "All Philosophers are Sophists." Thus "Only S is P" is equivalent to "All P is S" and *switches* the subject and predicate terms.

The logical relations that hold between pairs of categorical propositions can be mapped onto a square, known at the *Aristotelian Square of Opposition*.

There are three types of *oppositional* relations (indicated in red).

- CONTRADICTORIES: The propositions on *opposite diagonals* of the square are *contradictories*. The A and O, and the E and I, propositions always have *opposite* truth values.

For example, the contradictory of the proposition "All men are mortal" is "Some man is not mortal" and the contradictory of "No man is mortal" is "Some man is mortal." Exactly one of a pair of contradictory propositions is true and the other is false.

- CONTRARIES: The propositions on *opposite corners* of the *top edge* of the square are *contraries*–while the A and E propositions may both be *false*, it is *impossible* for both to be *true*, assuming that subject term isn't *vacuous* or *empty*.

For example, in contrast to contradictories, which always have opposite truth values, it is possible for the contraries to both be false. Suppose that men are typically mortal, but that some man is not mortal. Then

The best way to secure training and practice in such arguments is to get into the habit of transposing arguments. For in this way we shall be better able to deal with the matter at hand and shall thoroughly understand many arguments in the light of a few.
—ARISTOTLE, Topics viii.

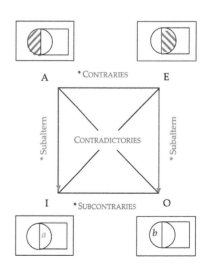

both "all men are mortal" and "all men are not mortal" are false. However, it cannot be the case that both "all men are mortal" and "no men are mortal" are true, assume there is at least one man, or that men exist.

- SUBCONTRARIES: The propositions on *opposite corners* of the *bottom edge* of the square are *subcontraries*–it is *impossible* for both to be *false*, assuming the subject term isn't *vacuous* or *empty*.

For example, at least one (possibly both) of the propositions "Some men are mortal" or "Some men are not mortals" must be true, assuming men exist.

Lastly, we have the logical relation known as *subalternation*, which represents *positive* implication and is indicated by the blue arrows on the sides of the square.

- SUBALTERNS: The universal affirmative A proposition *implies* the existential affirmative I proposition, and the universal negative E proposition *implies* the existential negative O proposition, assuming that the subject term represented by the circle isn't *vacuous* or *empty*.

The Venn diagrams that accompany the categorical propositions enable you to *visualize* these logical relations.

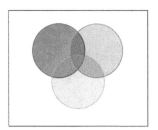

- If you superimpose the Venn diagrams for CONTRADICTORY propositions, you're forced to place a letter in an area that is shaded out. This is what a *contradiction* looks like. All the other relations have asterisks * as a reminder that these relations hold only if the subject terms are non-empty.

- Contrast the CONTRADICTORY relation with the CONTRARY relation that holds between the universal propositions. If you superimpose the Venn diagrams for the universal A and E propositions, you'll see the entire circle, representing the subject term, has been shaded out. If the subject term is non-empty, both A and E propositions cannot be simultaneously true. The A and E propositions are *contraries*.

- Consider the SUBCONTRARY relation. If you superimpose the Venn diagrams for the existential I and O propositions, you'll see that at least one of them (perhaps both) must be true, assuming the subject term to be non-empty. The I and O propositions are *subcontraries*.

- There are the two SUBALTERN relations. Given the A proposition is true and assuming the circle representing the subject term has a letter in it, you may infer the I proposition is true. Similarly, given the E proposition is true and assuming the circle representing the subject term has a letter in it, you may infer the O proposition is true.

With these logical relationships as background, let's turn to the evaluation of syllogisms. A *syllogism* is an argument consisting of two categorical premises and one categorical conclusion, having three terms each of which occurs twice. The *middle* term occurs once in each of the premises, and the two *end* terms occur one in the premises and once in the conclusion.

A syllogism is *valid* if the premises *imply* the conclusion or, alternatively, if it is *impossible* for the premises to be *true* while the conclusion is *false*.

In medieval times, students of logic memorized poems whose code letters would enumerate all the valid forms of the syllogism and learn a list of fallacies with impressive sounding names as:

Fallacy of the Undistributed Middle
Fallacy of the Illicit Major Term
Fallacy of the Illicit Minor Term
Fallacy of Exclusive Premises:
Fallacy of Two Negative Premises Implying a Positive Conclusion
Fallacy of Two Affirmative Premises Implying a Negative Conclusion

The traditional interpretation of the universal affirmative is somewhat different from the modern interpretation, which requires some fiddling with the O statement to make it contradictory to the A statement. Aristotle assumed that the A statement implies the I statement, e.g, that "All men are mortal" implies "Some man is mortal." This assumption is called "existential import". Modern logic does not assume existential import. For example, the warning 'All trespassers will be shot" does not imply that "Some trespasser will be shot." The modern interpretation has its own counterintuitive consequences. The modern interpretation assumes that universals whose subject term applies to nothing are "vacuously true." Thus, if there are no unicorns, "all unicorns have one horn" is (vacuously) true (but so is "all unicorns have two horns")!

Rather than memorizing such a complicated set of rules, we can determine the validity of a syllogism using just three rules and a geometric method of shading and lettering a Venn diagram. Both of the methods are based on construct the *reductio* set for a syllogism.

The *reductio set* for a syllogism is a set of propositions equivalent to the premises and the negation of the conclusion in *existential* or *negative existential* form. The *impossibility* of consistently representing the *reductio* set means that the original syllogism is *valid*. Consistently representing the *reductio* set *invalidates* the original syllogism.

The universal A and E propositions are equivalent to the *negations* of the O and I existential propositions, respectively and are diagonally across from them in the Aristotelian Square of Opposition. If diagonal

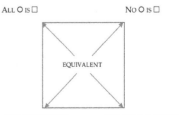

propositions are *contradictory*, then the negation of a proposition at one corner of the diagonal is equivalent to the proposition at the opposite corner.

The reason for expressing the propositions in the *reductio* set as positive or negative existentials is that they are intuitively easier to diagram. A negative existential is diagrammed by "shading out" the relevant area, and a positive existential is diagrammed by placing a letter in the relevant area—which we'll call "lettering in". A *contradiction* occurs when you're forced to place a letter in an area that is shaded out.

A syllogism is valid if it is impossible for its premises to be true while the conclusion is false. In terms of the *reductio* set, if the *reductio* set can be consistently represented, then the syllogism is invalid. If it is impossible for the *reductio* set to be consistently represented, then the syllogism is valid. For a syllogism to be valid, therefore, means that representing the *reductio* set forces one into a contradiction–i.e., forces one to place letter in an area that is shaded. Since contradictions must be forced, in terms of diagraming this means that you should represent negative existentials (by shading out) before positive existentials (by placing a letter in an area or on the borders of several areas). In other words, shade out before lettering in. Let's consider an example:

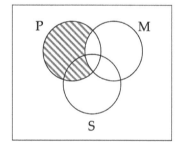

All philosophers are lovers of wisdom.
No lovers of wisdom are sophists.
∴ No sophist is a philosopher.

Paraphrase		Reductio Set	Venn
All P are M	≡	*Not*–(Some P is *not*–M)	
No M is S	≡	*Not*–(Some M is S)	
∴ No S is P	≠	Some S is P	*a*

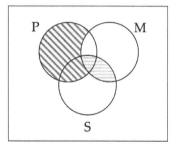

We represent the *negative existentials* by *shading out* the area of the Venn diagram corresponding to the area where the positive (unnegated) existential is true. In the above example, the negative existential corresponding to the first premise—*Not*-(Some S is *not*-M)—is represented by shading out the area in the S circle but outside of the M circle (this area shaded out in red below).

The second negative existential—*Not*-(Some M is P)—is represented by shading out the overlap between (or the *intersection* of) the M circle and the P circle.

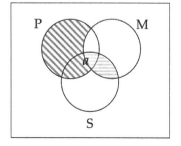

Finally, we represent the *positive existential*—*Some* S is P—by placing a letter in the intersection of the S and P circles. Here we place the letter 'a' on the border of the M circle to indicate it could have gone on either side.

We have been forced into a *contradiction*—i.e., placing a letter in a shaded out area. This means that it is impossible to represent the *reductio* set—i.e., it is *impossible* for the premises to be *true* while the *conclusion* of *false*. Therefore, the syllogism is *valid*.

To use the Venn diagrams to investigate the validity of a syllogism:

(V_1) Construct its *reductio* set by representing the premises and the opposite of the conclusion in existential or negative existential form.

(V_2) Represent negative existentials first by shading out the areas corresponding the negated positive existential. Then represent the (positive) existentials by placing a letter in the corresponding area(s) of the Venn diagram.

(V_3) A forced contradiction occurs when you must to place a letter in an area that has been shaded out. The impossibility of representing the *reductio* set is equivalent to the validity of the original syllogism.

Is there a *forced contradiction*? $\begin{cases} Yes & \text{the syllogism is } valid \\ No & \text{the syllogism is } invalid \end{cases}$

It happens there is a simple set of rules that can be applied to the *reductio* set to determine the validity of the original syllogism. These rules were set forth by the American logician and psychologist Christine Ladd-Franklin (1847 - 1930), who was a related to Benjamin Franklin.

According to the Ladd-Franklin rules, a syllogism is *valid* (in the modern sense) if and only if its *reductio* set simultaneously meets *all* of the following three conditions:

(LF_1) There is exactly *one positive existential*.

(LF_2) The pair of *negative existentials* has a contradictory *pair* of terms, i.e., a term in one that is negated in the other.

(LF_3) Each of the *terms* of the *positive existential* occurs *uniformly* positive or *uniformly* negative throughout the *reductio* set.

Failing any one of these rules renders the syllogism invalid.

Let's verify that our syllogism passes all three tests or rules. First, there is exactly one positive existential, the proposition provided by the negation of the conclusion. Secondly, the two negative existential propositions have the term M is common, which is positive in one and negative in the other. Thirdly, the terms of the positive existential, S and P each occur uniformly positive. Therefore, the argument, according to the Ladd-Franklin rules is valid.

Applying under the name "C. Ladd", Christine Ladd was offered a fellowship from Johns Hopkins, which tried to revoke the offer upon learning she was a woman. Ladd eventually took classes taught by C S. Pierce and wrote a dissertation "On the Algebra of Logic" with Pierce as thesis advisor.

		REDUCTIO SET	LADD-FRANKLIN	
All P are M	≡	*Not*-(Some P are *not*-M)	LF$_1$	Yes
No M is S	≡	*Not*-(Some M is P)	LF$_2$	Yes
No S is P	≠	Some S is P	LF$_3$	Yes

Let's work through another example:

> All philosophers are teachers.
> All sophists are teachers.
> ∴ Some philosophers are sophists.

This syllogism is invalid but we want to verify this by using both the Ladd-Franklin rules and the Venn diagrams based on its *reductio* set.

PARAPHRASE		REDUCTIO SET	LADD-FRANKLIN	VENN
All P are M	≡	*Not*-(Some S are *not*-M)	LF$_1$ - satisfied	⊘
Some S is M	≡	Some S is M	LF$_2$ - fails	a
∴ Some S is P	≠	*Not*-(Some S is P)	LF$_3$ - fails	○

Here LF$_1$ is satisfied because there is exactly one *positive existential*, the opposite of the conclusion. Also LF$_2$ is satisfied because the pair of *negative existentials* has a contradictory pair of terms, namely, M and not-M. Finally, LF$_3$ is also satisfied because the terms S occurs uniformly positive throughout the *reductio* set, and so does the term P occurs uniformly positive. (Notice we would also have had a different valid syllogism if the term P had occurred universally *negative* throughout the *reductio* set). According to the Ladd-Franklin rules, the syllogism is valid.

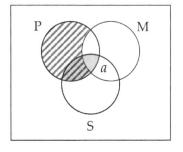

This concludes our brief introduction to the Aristotelian syllogism, which avoids such confusing doctrines as "distribution" and controversies as "existential import." These issues are left for a second module of *Thinking Matters* which embeds the theory of the syllogism in a more general logical setting. It is sufficient for now to have easy to understand methods that solve the syllogism in ways far less baroque than traditional treatments, which are quaint but antiquated.

Rhetoric & Persuasion

Rhetoric is the study of the techniques of persuasive argumentation. Logic, on the other hand, is the study of correct argumentation. The rhetorician evaluates an argument from the point of view of its persuasiveness: how effectively did the author of the argument persuade his audience? The logician looks at the argument from the point of view of its correctness: did the premises of the argument really support or provide a justification for the conclusion that was drawn? Not all rhetorically persuasive arguments are logically correct, and not all logically correct argument are rhetorically persuasive.

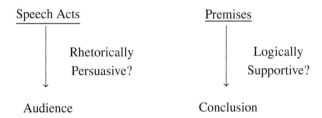

Rhetorical and Logical Arguments

An argument from a rhetorical point of view is a series of speech acts aimed at persuading an audience. An argument from a logical point of view is a series of premises offered in support or justification of a conclusion.

Although there has been a history of antagonism between philosophers and rhetoricians since the days of ancient Greece, each discipline without the other is sterile. Truth without persuasiveness is futile; persuasiveness without truth is blind. According to the famous statesman-philosopher Cicero (106-43 BC), the highest human achievement lies in the effective use of knowledge for the guidance of human affairs.

In the second module we'll learn the art of argumentation. An argument, as the logician uses the term, is a series of statements (called premises) offered in support or justification of another statement (called the conclusion). Some argument forms occur with such frequency in argumentative discourse that they've acquired traditional names. One such argument form is known as an analogy.

A famous analogy was used early in the abortion debate by M.I.T. Philosopher Judith Jarvis Thomson. In her article, "A Defense of Abortion," Thomson argues against the reasoning that abortion violates the 'right to life' of the fetus. She argues that even if we grant that a fetus is a person, abortion can be morally permissible. She uses several analogies and thought experiments in her argument. One of them follows:

> ...But now let me ask you to imagine this. You wake up in the morning and find yourself back to back in bed with an unconscious violinist. A famous unconscious violinist. He has been found to have a fatal kidney ailment, and the Society of Music Lovers has canvassed all the available medical records and found that you alone have the right blood type to help. They have therefore kidnapped you, and last night the violinist's

circulatory system was plugged into yours, so that your kidneys can be used to extract poisons from his blood as well as your own. The director of the hospital now tells you, 'Look, we're sorry the Society of Music Lovers did this to you–we would never have permitted it if we had known. But still, they did it, and the violinist now is plugged into you. To unplug you would be to kill him. But never mind, it's only for nine months. By then he will have recovered from his ailment and can be safely unplugged from you.' Is it morally incumbent on you to accede to this situation? [5]

Do you see the point of Thomson's analogy? Thomson believes that you are not morally obligated to remain plugged into the violinist because he has no right to use your body. Even if we assume, as Thomson does, that to unplug the violinist is to kill him rather than merely to withhold life-support, she claims that it is permissible for you to unplug.

For Thomson the central issue in determining the moral permissibility of abortion is not whether the fetus is a human being but whether a fetus has a right to the use of the mother's body. For the sake of the analogy Thomson grants that a fetus is a full-fledged human being. However, Thomson argues that a fetus has a right to the mother's body only if the mother has assumed a special responsibility towards the fetus either implicitly or explicitly. Thomson argues that no special responsibility is assumed by the woman in many cases of abortion–for example, in cases of rape or in cases in which the parents have taken all reasonable contraceptive precautions to prevent conception or in cases of conception due to biological ignorance. Thomson holds this view because, unlike the 'right to life' position, it "allows for and supports our sense that, for example, a sick and desperately frightened fourteen-year-old schoolgirl, pregnant owing to rape, may of course choose abortion." What are your thoughts on this analogy? Is it persuasive?

In the argument form of *analogy*, similarities between two cases are used as a basis for inferring a further similarity. To get an idea of how we can systematically approach the analysis of such arguments on emotionally charged issues, let's consider a simpler one.

When George H.W. Bush was Vice President in the Reagan administration, he was criticized by the press for not voicing his pre-primary opposition to President Reagan's economic policies. Bush defended himself by stating, "A player on a football team does not tackle his own quarterback." Now Bush's reply can be construed as an argument by analogy.

Represented schematically, an analogy has the following form:

(1) X IS SIMILAR TO Y
(2) Y HAS PROPERTY F (*because of generalization* G)
(3) SO, X HAS PROPERTY F.

The premises in lines (1) and (2) give reasons in support of the conclusion in line (3).

Analogy is perhaps the most frequently used argument form in both legal and moral reasoning. We shall therefore adopt some useful terminology which may remind us of this fact. In the above schema, we shall say X represents the controversial case, Y the precedent case. We shall refer to the property F as the ruling in the precedent case. The letter 'G' stands for a principle of generalization which justifies the ruling reached in the precedent case.

We may now set forth, or explicate, Bush's analogy by paraphrasing it into the above standard form. We all have a tendency to criticize the arguments of others before we really understand what they've asserted. It is therefore useful to cultivate the habit of explicating—or clearly setting forth—an argument before proceeding to evaluate it.

In explicating the analogy, some notation will be useful. We'll explicitly list the respects in which the precedent and controversial cases are similar by placing a double arrow $\iff$ between the similarities. This will remind us of specific ways in which the analogy can be challenged. Since analogies often crucially depend on relational similarities, we shall use the notation '$x : y$' to represent a relationship that exists between x and y. Stating Bush's argument in the above schematic format we obtain:

1. The presidential administrative team is similar to a football team.

President	$\iff$	quarterback
Vice President	$\iff$	player on team
Vice President : President	$\iff$	player : quarterback
criticizing the President	$\iff$	player tackling own quarterback

2. A player on a football team shouldn't tackle his own quarterback (because the team's success depends on each player following the game plan).

3. So, a Vice President should not criticize the President.

Having set forth or explicated the argument, we can now proceed to evaluate it. Three fallacies, or errors in reasoning, are frequently committed in arguments by analogy.

- A *Weak* or *False Analogy* occurs when the analogy draws on false, irrelevant or superficial similarities or the analogy overlooks relevant dissimilarities.

Ronald Reagan became the 40th U.S. President with George H. W. Bush as his running mate. The political slogan "win one for the Gipper" was later used by Reagan, alluding to his portrayal of Frank Gipp in *Knute Kockne: All American* [1940].

For example, if Bush were to have offered the analogy solely on the basis of Ronald Reagan's performance as George Gipp in the movie *Knute Rockne: All American*, he would be basing his analogy on an irrelevant, accidental similarity. However, there is no reason to fault Bush on this account. His point is that Reagan as President is the 'quarterback' or leader of the presidential team. This point is not diminished by the superficial coincidence provided by Ronald Reagan's earlier career as a movie actor.

A good analogy also should not overlook relevant dissimilarities. The controversial case need not be exactly like the precedent case. An administration, for instance, is made up of thousands of people, whereas a football team involves, say, about 40 people. But this dissimilarity does not diminish the point of Bush's analogy, which is a point about what is required by teamwork, no matter what the actual size of the team.

There should not be any obvious differences between the two cases that affect the drawing of the analogous ruling or conclusion. The purpose of a presidential administration is, for example, importantly different from that of a football team. The purpose of a football team is to win the game or to provide entertainment for the spectators. The purpose of the presidential team is supposedly not merely to win the next election or to entertain the citizens. Hopefully, its purpose is to draft intelligent and wise policies. An open discussion in the football huddle would be wasting valuable time and undermine the unity and effectiveness of the team. On the other hand, drafting wise policies may require the airing and debating of different opinions on a given issue. Even if openly criticizing the President in the press is inappropriate, it might have been appropriate for Bush to air some constructive criticism in private.

- The fallacy of the *Faulty Precedent* occurs when even the ruling in the precedent case can be called into question.

If the precedent case is the basis for drawing a conclusion in the controversial case, the ruling in the precedent case must not be beyond serious dispute. So we might ask, "Are there ever situations in which it is appropriate for a team member to tackle his own quarterback?" Suppose the quarterback is running in a confused daze toward the opposition's goal. In such cases, it may be perfectly appropriate for a player to tackle his own quarterback. (In raising such a possibility, we are not implying that President Reagan actually led the country in a confused daze towards the opposition's goal line!)

- The fallacy of **Faulty Generalization** occurs when the principle of generalization—which may be explicit or implicit—is untrue, unclear, or subject to significant doubt.

Is it really true that teamwork requires the non-dissenting obedience of its members? Perhaps teamwork requires more and less. A good volleyball team, for example, does not merely play according to rigidly preconceived roles, but may alter its strategy to accommodate the unexpected cases in which blind obedience on the part of its members would lead to a less effective team. On the other hand, if the principle of generalization were carefully restricted so that it would be true in the football case, it is not at all clear that the same generalization would still support the analogous conclusion to be drawn about presidential administrations.

In a later section, we will carefully explicate Judith Jarvis Thomson's argument in this fashion. Have any of the three fallacies associated with analogy been committed in her argument?

Science or Pseudoscience?

In 1695 Edmond Halley began applying Newton's ideas to the motions of comets. Observing a comet in 1682, Halley began researching the previous sightings of comets in the same location relative to the positions of the sun and the stars. He found two previous recorded observations of comets in approximately the same location in 1606 - 1607 and in 1530 - 1531. Assuming these observations to be of the same comet, Halley calculated that the comet would reappear on Christmas Day in 1758. Halley died fifteen years before the comet's return, but his prediction was remembered and when it was confirmed, the comet was officially named "Halley's comet." This prediction was in fact decisive in swaying many scientists to accept the previously controversial Newtonian model of celestial mechanics.

Compare the above case of scientific prediction with the astrological prediction of Mark Twain. "I came in with Halley's comet in 1835. It is coming next year, and I expect to go out with it", Mark Twain told his biographer, Albert Bigelow Pain. "The Almighty has said, no doubt: 'Now here are these two unaccountable freaks; they came in together, they must go out together.'" Mark Twain, his biographer noted, wasn't disappointed. His death, on April 21, 1910, came one day after Halley's comet.

Is there a difference between these two cases? Why do we normally suppose that the success of Halley's one prediction provides substantial justification for Newton's theory of celestial mechanics, whereas the success of Mark Twain's prediction offers little support for the claims of astrology? Is there any logical basis for such a supposition or is it mere prejudice?

Alphabetology

In Module IV we'll compare the conditions for a well-designed scientific test with the fallacies of theory testing often used to support the claims of what has been called 'pseudoscience.' To stimulate your thinking about these matters, consider the case of *Alphabetology*. Alphabetology was invented by Clark Glymour and Douglas Stalker in their humorous article "Winning Through Pseudoscience."[6]

To make an Alphabetological reading for your last name, you need to consult three charts. These charts help you to find your Vowel Sign, your Ascending Consonant Group, and your Syllable House.

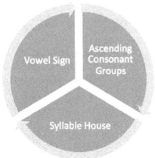

VOWEL SIGN
A sincere, strong, patient
E lucky, fluid, engaging
I proud, forceful, possible born leader
O aggressive, impatient, fighting instinct
U gloomy, scholarly, punctual
Y unpredictable, skillful, natural executive

ASCENDING CONSONANT
Group 1: b,c,d,f,g benefits, advantages
Group 2: h,j,k,l,m conflicts, strife
Group 3: n,p,q,r,s difficulties, deviousness
Group 4: t,v,w,x,z resolutions, conclusions

SYLLABLE HOUSE
1 syllable commutative (combining independent of order)
2 syllables distributive (diffusing more or less evenly or unevenly)
3 syllables associative (relating by groups)
4 syllables symmetric (having corresponding yet connecting points)
5 syllables contrapositive (permutating major parts, probably not positive)

Form for your very own Alphabetological reading.

READING FOR:	(your last name)
VOWEL SIGNS:	(list vowels in lastname in a column)
ASCENDANT:	(list first consonant in last name)
HOUSE:	(number of syllables in last name)

For example, here's Stalker's chart for the Alphabetological reading for Richard Nixon:[7]

READING FOR:		NIXON
VOWEL SIGNS:	*i*	proud, forceful, born leader
	o	aggressive, impatient, fighting instinct
ASCENDANT:	*n*	difficulties, deviousness
HOUSE:	2	distributive

The interpretations for each of the letters are obtained from the corresponding charts in the obvious way. This is then the basis for constructing a reading.

Stalker, tongue in cheek, warns us that "our principles cannot be reduced to mechanical operations" and that "the human factor is ever present and irreducible here, as it is throughout our real lives and activities, in contrast to the fictional world scientists describe for us." For example, given the above chart for Nixon, Stalker gives the following reading:

> Nixon is certainly a proud person, as his public appearances and writings clearly indicate. He is also forceful, aggressive, and impatient. These traits were shown time and time again in national and international politics. Nixon, more than most world figures, has a fighting instinct. After all, isn't he still fighting? His Ascending Consonant points to the Watergate affair, among many lesser problems. His Signs and Ascendant seem to distribute evenly as can be seen even by the untutored eye. Nixon, devious? The reading tells all. So be it! [8]

The reading, claims Stalker, is so accurate that it is hard to believe. To show that the reading is not a fluke, he gives another quick and accurate sample reading. This time the reading is for Adolf Hitler.[9]

From this chart Stalker gets the following reading:

READING FOR:		HITLER
VOWEL SIGNS:	*i*	proud, forceful born leader
	e	lucky, fluid, engaging
ASCENDANT:	*h*	conflicts, strife
HOUSE:	2	distributive

> Although infamous, Hitler was admittedly proud, forceful, and engaging—a born leader with more than his share of luck to be at the right time to take over a mighty nation.
>
> Indeed, an amazingly lucky chain of events that brought Hitler to power, one only a very lucky man could chance on—from housepainter to Führer; think of it! His changing views and whims and fancies showed him to be fluid in his preferences, most often to his discredit and the dismay of others. One week you could be his friend, the next his enemy. Certainly his life was marked with conflicts, so much so that his traits distributed in the direction of definite evil. An uneven total, regrettably. One can only wonder what he might have been if they had coalesced differently.[10]

Let's try an alphabetological reading for our own time:

Reading for:		TRUMP
Vowel Signs:	u	gloom, scholarly, punctual
Ascendant:	t	resolutions, conclusions
House:	1	commutative

The vowel sign for TRUMP is U, which stands for "gloomy, scholarly and punctual", but this doesn't fit the person at all, as we know him. What's wrong?

It turns out that the actual surname of the Trump family was DRUMPF. When this kind of obfuscation occurs, the Vowel Sign is tossed out and instead of considering the fake Ascendant Consonant; the reading takes into account *all* the consonants of the *actual* family name. When this is done we have the following consonant groups:

Group 1:	b, c, D, F, g	benefits, advantages
Group 2:	b, h, j, l, M	conflicts, strife
Group 3:	n, P, q, R, s	difficulties, deviousness

Actual Alphabetological Reading: More than any previous President, this President has used the Office for *"benefits and advantage"*–in fact, you can even say he has "doubled down" (double consonants in group 1) benefiting himself and his confederates. No matter what your political persuasion, everyone agrees that the tone of the nation has certainly been one of *"conflicts and strife."* Finally, this President's strategy has been never to admit mistakes but to "double down" (double consonants in groups 3) producing *"difficulties and deviousness."* While initially appearing to refute Alphabetology, this reading confirms its accuracy!

Have you ever found astrological forecasts or handwriting analyses or the like to be astoundingly accurate?

Probability, Statistics & Decision Making

"Probability is the very guide of life," said Bishop Joseph Butler and others before him (such as the Roman orator Cicero). The basic concepts of probability are simple, yet this branch of reasoning swarms with paradoxes and counterintuitive results.

Consider the following paradox known as **Simpson's paradox**. A study in 1973 at the University of California investigated possible sex bias in graduate school admissions. About 44% of men applying for graduate work were admitted, yet only 35% of women were admitted.

The qualifications of the men and women were roughly the same, so this seemed to be a clear case of sex discrimination. However, when the same data were examined to determine which departments were responsible for the discrimination, it was discovered that in each department women had a greater chance of being accepted than men.[11]

	Applicants	% admitted
Men	8442	44%
Women	4321	35%

When examining the individual departments, it was found that no department was significantly biased against women, in fact, most departments had a small bias against men.

Major	Men		Women	
	Applicants	% admitted	Applicants	% admitted
A	825	62%	108	82%
B	560	63%	25	68%
C	325	37%	593	34%
D	417	33%	375	35%
E	191	28%	393	24%
F	272	6%	341	7%

How can this paradox be explained?

Consider another problem which became newsworthy. Marilyn vos Savant publishes a column entitled "Ask Marilyn." She is touted to be the individual listed for "Highest I.Q." in the Guinness Book of World Records. In September of 1990, she published the following problem known as "Monty's Dilemma."

Monty Hall, a thoroughly honest game-show host, has placed a car behind one of three doors. There is a goat behind each of the other doors. "First you point to a door," he says, "then I'll open one of the other doors to reveal a goat. After I've shown you the goat, you make your final choice and you win whatever is behind that door."

You want the car very badly. You point to a door. Hall opens another door to show a goat. There are two closed doors remaining, and you have to make your decision: should you stick with your first choice or should you switch to the other unopened door? Or doesn't it matter?

Vos Savant said it was better to switch. After giving her answer, she received over 10,000 letters, the great majority disagreeing with her. Some of the most vehement criticism came from mathematicians who alternated between gloating at her ("You are the goat!") and lamenting the nation's innumeracy ("As a professional mathematician, I'm very concerned with the general public's lack of mathematical skills. Please help by confessing your error and, in the future, being more careful.")

What should you do and why?

Rather than tackle this tantalizing puzzle now, let us again consider a simpler probability problem which will illustrate a model, known as a probability tree, which can be used to solve such problems.

Eve's Children

Eve says, "Of my two children at least one is a boy." Given this information and assuming that the probability of having a boy is 1/2, what is the probability that both of Eve's children are boys?

(A) 1/4
(B) 1/3
(C) 1/2
(D) 3/4
(E) None of the above.

Probability swarms with plausible but specious lines of reasoning. For example, given the above problem, many people reason as follows:

> "We know that one child is a boy. Hence the remaining child is either a girl or a boy, and so there is a 50-50 chance of two boys. Therefore the probability of two boys is 1/2."

Others disagree and reason as follows:

> "There are four options–Boy-Boy, Boy-Girl, Girl-Boy, and Girl-Girl. Only one of the four options is Boy-Boy. The probability of two boys is 1/4."

Unfortunately, both pieces of reasoning are fallacious. Can you spot the fallacies?

Later we'll investigate the different senses of 'probability' or 'chance'. For example, when we say that hospital records show that boys are actually more probable than girls, we are using a statistical or frequency interpretation of probability. When we say the chances of rolling double sixes on a pair of fair dice is 1/36, we are using 'probability' in the *a priori* or classical sense. According to this latter concept of probability, the probability of an event is the ratio of the number of ways the event can occur over the total number of ways, where all the ways are equally likely.

$$\text{Probability (Event)} = \frac{\text{\# of favorable ways}}{\text{total \# of equally likely ways}}$$

Before solving the problem analytically, let's simulate the problem. Let's say heads represents Eve's having a boy and tails having a girl.

Our simulation will involve flipping two coins—a penny and a nickel, representing the sex of the first and second child, respectively.

Record how many times, you two boys (two heads) and how any time we get one boy and one girl (one head and one tail). Since the outcome of two girls (two tails) is ruled out by the information given in the puzzle, if you flip this outcome, ignore the result and repeat the experiment until you get a result that is acceptable according to the conditions of the puzzle. Do this until you get 30 acceptable outcomes.

Tally your results in a table:

2 boys	1 boy and 1 girl
‖‖‖‖	‖‖‖‖ ‖‖‖

Is your simulation closer to 25%, 33% or 50%?

Record your results for 30 trials. How many times did you get a double heads? Is the ratio of number of double heads to the total number of trials closest to $1/4$, $1/3$, or $1/2$? If it's closer to $1/3$, does that prove the answer is $1/3$?

Now let's solve the problem analytically and model is using a probability tree.

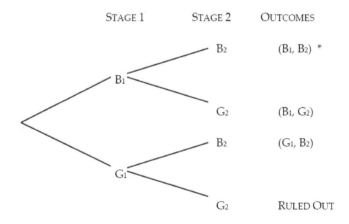

To solve our puzzle, let's first describe the entire 'experiment' of having children as a sequence of stages. We'll order the stages using the words 'and then. In this example, Eve had her first child and then she had another. Next we'll describe the outcome of each of the stages of the experiment. We'll state the outcomes using the words 'or else'. In our example, Eve's first child was either a boy or else a girl. The outcomes for the second stage depend on the first. If Eve's first child was a boy, then her second child was either of boy or else a girl. If her first child was a girl, on the other hand, then her second child must be a boy. We know that at least one of the two children is a boy. We may graphically represent these outcomes in a probability tree.

A probability tree diagram is a convenient way of visualizing all the possible outcomes of our two stage experiment. Beginning with the origin at the left, the tree splits into branches at the various nodes and terminates at the endpoints. A path is a connected series of branches starting at the origin and stopping at an endpoint. Each path represents a possible outcome of the entire chance experiment.

Given the tree diagram for our problem, we can easily see that the total number of equally likely outcomes of the experiment is 3. Of these, only one is the outcome where both are boys (this possibility is indicated by the '*'). Consequently, the answer to our problem is:

$$\text{Probability } (B_1, B_2) = \frac{\text{\# of favorable ways}}{\text{total \# of equally likely ways}} = \frac{1}{3}$$

What is wrong with the original fallacious arguments? Notice that if we were to assume that the *first child is definitely a boy*, then the probability of two boys is 1/2. In such a case, we would only be dealing with the upper two branches of the tree. On the other hand, if we were to assume the *second child is definitely a boy*, then the probability of two boys is also 1/2. The second branch with the second child a girl would be pruned from the tree of possibilities.

However, in the original problem we are only given that *at least one* of the two children is a boy. So, we can arrive at the outcome of one boy and one girl in two possible ways–by having the boy first and then the girl, or having the girl first and then the boy. The first piece of fallacious reasoning that arrives at the answer 1/2 overlooks this fact. And the second piece of fallacious reasoning simply overlooks the fact that the possibility of two girls is explicitly ruled out by the stated conditions of the puzzle.

In our module on scientific reasoning we will strive to gain a conceptual understanding of the basic concepts of statistics, probability, and decision making, (as opposed to merely computational).

Newcomb's Paradox

Here is a tantalizing problem known as *Newcomb's Paradox*,[12] named after the physicist William Newcomb, who invented it. Newcomb's paradox involves a Being who has the ability to predict your choices with almost complete accuracy. The Being has predicted your choices in many other situations and the choices of many other persons in the situation to be described. Let's say that the Being is accurate, say, 90% of the time.

You are to play a game with the Being. You come into a room in which there are two boxes–A and B. Box A always contains $1000, and box B either contains $1 million or nothing. You can choose to take both box A and box B or to take box B alone. If that were all there were to the puzzle, of course, you should take both boxes.

But there's an intriguing complication. The Being will put either $1 million or nothing in box B based on the Being's prediction about how you will choose. If the Being predicts that you will choose box B alone, the Being will, so to speak, reward your faith by putting the $1 million in box B. If, on the other hand, the Being predicts you will choose both boxes, the Being will put no money in box B. (Furthermore, if the Being predicts that you will base your choice on some random event like flipping a coin, the Being still won't put any money in box B.)

You know all these facts, the Being knows all these facts, and you both know that the other knows these facts. The Being makes his prediction and puts either the $1 million or nothing in box B. What should you do?

Some argue that choosing box B alone is the more rational choice. They argue that there's a 90% chance the Being correctly will predict my choice. If I choose box B alone, there's a 90% chance the Being will have predicted this, and so I'll have a 90% chance of getting $1 million. If I choose both boxes, there's a 90% chance the Being will have predicted this, and so I'll have a 90% chance of getting only $1000. Since 90% of $1 million is a lot larger than 90% of $1000, I should choose box B alone. This is called the "Expected Value" argument and is modeled by the following payoff matrix:

Matrix for the Expected Value Argument

	Correct Prediction	Incorrect Prediction
Take Both Boxes	$1,000	$1,001,000
Take Box B Alone	$1,000,000	$0

To calculate the respective expected values more precisely:

E.V. (Both)   = 90% ($1,000)   + 10% ($1,001,000) = $101,000
E.V. (B Alone) = 90% ($1,000,000) + 10% ($0)   = $900,000

The choice of taking both boxes is always better—"dominates"—choosing box B alone. This is called the "Dominance" argument and is modeled by the following payoff matrix:

Matrix for the Dominance Argument

	$1M is in Box B	$1M is Not in Box B
Take Both Boxes	$1,001,000	$1,000
Take Box B Alone	$1,000,000	$0

Taking both boxes is always better than (i.e., *dominates*) choosing B alone no matter what.

To solve Newcomb's paradox it is not sufficient to merely affirm one of these arguments. Both positions have a persuasive payoff matrix and an associated argument. Instead you must undermine the opposing argument and explain why, for example, it contains a false assumption. Would the solution to th puzzle change if the game were to be iterated and played ten times in the row? As we shall see in a future module, Newcomb's paradox is related to another famous paradox in game theory known as the Prisoner's dilemma.

A Parting Paradox

Protagoras

In ancient Greece there existed a class of teachers who came to be known as sophists. They traveled from city to city, much like wandering minstrels, and for a fee these sophists would teach their students how to speak persuasively on many different kinds of topics. The following story, known as the Retort, concerns Protagoras (481-411 BC), one of the greatest of the sophists.

Protagoras had agreed to teach logic to Euathlus. They decided that Euathlus would have to pay for lessons if he won his first case in court, and that he would not have to pay otherwise. After receiving his lessons, however, Euathlus delayed taking any cases in court.

Protagoras, worrying about his reputation as well as wanting the money, decided to sue. In court Euathlus, who had been a good student argued thus:

> Your honor, either I win this case, my first appearance in court, or not. If I win, I will not have to pay for I will have won the suit for payment. If I lose, I will have lost my first case in court so I will not have to pay by the terms of our contract. So either way, I will not have to pay.

But Protagoras, the consummate sophist, gave the following retort:

> "Your honor, Euathlus is right, either he wins this case or he does not. If he wins, then he must pay by the terms of our contract since he will have

won his first case in court. If he loses, then he must pay because he has lost the suit for payment. So, your honor, either way, Euathlus must pay.

Who should have won the case? It is clear that in order to sort out such paradoxes, we shall need to inquire into the general theory of validity. Later, we will investigate the notion of validity and develop an intuitively natural formal system of deduction that will enable us to see through an argument to its underlying logical (or illogical) structure. With this system of rules and strategies, you will be able to construct such arguments as those above with more fluency and to criticize such arguments with more insight.

The collection of models introduced in this chapter are not meant to be comprehensive. They are meant to be comprehensible, a toolbox of flexible thinking tools. Moreover the puzzles and paradoxes posed in this chapter, in addition to having a certain intrinsic charm of their own, are intended to be catalysts to reconnect you to the joy of problem solving.

"The most incomprehensible thing about the universe is that it is comprehensible."
—Albert Einstein

Summary of Concepts

In this chapter, we illustrated different types of models. These models provided us with a preview of later developments.

A Matrix is an array of boxes that can be used to represent deductions by a process of elimination.

A Truth Table is a method for analyzing a complex statement in terms of an exhaustive list of possible assignments to its simpler component statements. Truth tables can be used to solve truth-teller/liar puzzles.

A Directed Graph is a pictorial way of displaying a system of dependent relations. Such a graph consists of connecting or "grafting" together four types of T-bars. Directed graphs can be used to solve sequencing problems and deductions.

Argument Schemata represent common patterns of argument. Argument schemata for conditional arguments include the valid argument forms of *modus ponens* and *modus tollens* and the associated fallacies of *affirming the consequent* and *denying the antecedent*.

Another common argument form is ANALOGY.

(1) X IS SIMILAR TO Y
(2) Y HAS PROPERTY F (BECAUSE OF GENERALIZATION G)
(3) SO, X HAS PROPERTY F.

Three fallacies are associated with the argument form of analogy.

1. The fallacy of WEAK or FALSE ANALOGY occurs when the analogy draws on false, irrelevant or superficial similarities, or overlooks relevant dissimilarities.
2. The fallacy of FAULTY PRECEDENT occurs when even the ruling in the precedent case can be called into question.
3. The fallacy of HASTY GENERALIZATION occurs when the principle of generalization, which may be explicit or implicit, can be called into question.

A SYLLOGISM is an argument consisting of two categorical premises and one categorical conclusion, having three terms each of which occurs twice. The middle term occurs once in each of the premises, and the two end terms occur one in the premises and once in the conclusion.

A SORITES is a chain of syllogistic arguments with missing intermediate conclusions. Arguments with implicit premises or implicit conclusions are called *enthymemes*.

The validity of a syllogism can be easily determined on the basis of its REDUCTIO SET, a set of propositions, in existential or negative existential form, equivalent to the premises and to the opposite of the conclusion. A syllogism is valid if and only if it is impossible to represent its *reductio* set consistently.

To use the VENN DIAGRAM method, represent negative existentials in the *reductio* set first by "shading out" and then represent positive existentials by placing a letter in the relevant (non-shaded) area. The *reductio* set is inconsistent if there is a forced contradiction, i.e., you are forced to place a letter in a shaded area. In such a case, the syllogism is valid, and it is invalid otherwise.

A syllogism is valid if and only if it passes all three of the LADD-FRANKLIN RULES for the *reductio* set:

LF$_1$ There is exactly one POSITIVE existential proposition in the *reductio* set.
LF$_2$ The negative existential propositions have a CONTRADICTORY pair of terms.
LF$_3$ Each of the terms of the positive existential occurs either UNIFORMLY *positive* or UNIFORMLY *negative* throughout the *reductio* set.

A PROBABILITY TREE is a graphic way of enumerating the outcomes of a chance experiment. To construct a probability tree :

1. Describe the STAGES of the experiment using the words "AND THEN."
2. Describe the possible OUTCOMES at each stage of the experiment using the words "OR ELSE." The OUTCOMES of each stage of the experiment are represented by a Fan of branches.
3. Each PATH in the probability tree represents an OUTCOME of the entire experiment.

Assuming that the outcomes at each of the vertices are equally likely, we may compute the probability of an event as follows:

$$\text{Probability (Event)} = \frac{\text{\# of favorable ways}}{\text{total \# of equally likely ways}}$$

In a later module, we will learn ways to prune and simplify these trees by decorating them with numbers and using the laws of probability.

Exercises

Group I

1. Do You Hobbit?

You can return to the Shire of Three Farthing Stone if you have either both the services of Bilbo Baggins and Strider or both the approval of the Council of Elrond and an Elvish cap. A wish from Gandalf will get you either the services of Bilbo Baggins or an Elvish cap but not necessarily both (and you don't know which). If you have the Ring of Power you can obtain both the services of Strider and the approval of the Council of Elrond. Outwitting the Dark Riders and answering the riddle of the Golem will get you the Ring of Power. You want to return to the Shire of Three Farthing Stone to enjoy the life of a retired thief. Which of the following (if any) will guarantee that retirement?

(A) outwitting the Dark Riders and obtaining the Ring of Power;
(B) answering the riddle of the Golem and outwitting the Dark Riders;
(C) obtaining a wish from Gandalf and the services of Bilbo Baggins;
(D) obtaining both the services of Strider and a wish from Gandalf;
(E) obtaining a wish from Gandalf and the Ring of Power.

2. Lifeboat Ethics

The great 19th century case of Dudley v. Stephens shocked Victorian sensibilities when three British sailors were convicted of cannibalism and murder. After their ship went down, the three British sailors and an ailing cabin boy were adrift in the lifeboat without food or water. From the grisly facts, it appeared that the trio stabbed to death the ailing cabin boy and drank his blood. The cabin boy, who was seriously ill to begin with, could not reasonably have been expected to live more than a couple of days. On the following day, the lifeboat was sighted by a passing ship and the survivors rescued.

The defense counsel pleaded that the action of the defendants was morally permissible. Indeed, he argued it was even morally obligatory that one person be sacrificed to save the lives of the others. The defense counsel likened the case to a situation in which three mountaineers, supported by one thin rope, find themselves dangling perilously over a cliff. It is evident to each of the mountaineers that unless one of them is cut loose, the rope will break and all will be killed. Since it is better to save two lives than to lose three, the two mountaineers at the top of the rope are morally justified in cutting loose the mountaineer dangling at the end of the rope.

Explicitly set forth the counsel's argument using the schema for argument by analogy. Then critically evaluate the argument by briefly explaining your answers to the following questions:

(A) Does it commit the fallacy of weak analogy?
(B) Is it based on a faulty precedent case?
(C) Does the argument rest on a faulty generalization?

3. *Knights and Knaves*

The inhabitants of Crete belong to one of two clubs–the Knights or the Knaves. Knights always tell the truth, but Knaves always make false statements. Post and Quine are two inhabitants.

P: Post is a Knight
Q: Quine is a Knight.

Let's say that Post asserts a proposition S. In general, we can't determine whether what Post said is true or false. However, we do know that Post is a Knight if and only if what he said is true. That is, we know that the following biconditional statement is true:

$$P \leftrightarrow S.$$

These types of facts are key to solving these logic puzzles. After deducing your answer by your own means, check it by constructing a truth table.

1. Given the following scenarios, to which club does Post and Quine belong?
    (A) Suppose Post says, "We are both Knaves."
    (B) Suppose Post says, "At least one of us is a Knave."
    (C) Suppose Post says, "If I am a Knight, then so is Quine."
    (D) Suppose Post says, "I am a Knight if and only if Quine is."
    (E) Post says, "If I am a Knight, then Lance loves Gwen."
    Prove that Lance loves Gwen!
2. This time you are talking to three inhabitants of Wonderland–Art, Gwen, and Lance.
    (A) Art says, "Gwen is a Knave."
    (B) Gwen says, "Art and Lance are of the same type."

Inhabitants are of the *same type* if they're both Knights or they're both Knaves. What is Lance?

4. *Breaking the Code*

Four espionage agents—Emil, Alan, Alonzo, and Norbert—are looking for the code book. There is one real code book and three fake code books. The code books are each written in one of the following four languages—French, German, Italian or Chinese—and no two books are written in the same language. Furthermore, each book is located in a different one of the four cities mentioned below. You know also the following facts:

(A) Emil found the Italian code book in Geneva;
(B) the agent who found the French code book did not find it in Bonn;
(C) the real code book was not found in either London or Peking, but fake code books were found there;
(D) Alan found a code book in Bonn;
(E) Alonzo found the Chinese code book, but he correctly deduced that it was fake;
(F) Norbert did not find his code book in Peking;
(G) Emil did not find the real code book.

Which agent found the real code book, in what language was the code book written, and in what city was it found?

5. *Syllogistic Adventures*

Determine whether the following syllogisms are valid or invalid using both Venn diagrams and the Ladd-Franklin rules for the *reductio* set.

(A) Some Hobbits love adventures.
All heroes love adventures.
∴ Some Hobbits are heroes.

(B) All Hobbits love creature comforts.
None who love creature comforts have adventures.
∴ No Hobbits have adventures.

(C) Only Hobbits wear bright colors.
All who love teatime wear bright colors.
∴ All Hobbits love teatime.

(D) All Elves have leaf-shaped ears.

Not all who have leaf-shaped ears are ambidextrous.

∴ Some Elves are ambidextrous.

(E) Only Orcs are obscene.

None who are obscene are obsequious.

∴ No Orcs are obsequious.

6. *Spies and Ballerinas*

There are six women in a Russian dance troupe. Four of them are real ballerinas and two of them are actually spies. Choose a pair of women at random. What is the probability that there is at least one spy among the pair?

(A) Make a guess before computing the answer.
(B) Simulate the problem using a deck of cards.
(C) Use a probability tree to find the answer. (Do you see a way of consolidating similar branches and calculating the answer by weighting the branches?)

Group II

7. *Mating Sequence*

You don't have to know how to play chess to solve this problem. (In fact, it's better if you don't think of it as a real game at all.) If Bobby sacrifices his knight [SAC-KN], then Boris will either capture the knight with his bishop [B × KN] or move the pawn in front of his king [KP]. If Boris captures the knight with his bishop, then his bishop will be removed from guarding his king [REM-B] and Bobby can move his Queen to KN7 [Q-KN7]. If Boris moves the pawn in front of his king, then he exposes his KN2 and Bobby can move his Queen to KN7. If Boris exposes his KN2, then Bobby's bishop can attack the long diagonal [B-DIAG]. Bobby threatens mate on the diagonal [MATE-DIAG] if he can both move his Queen to KN7 and also have his bishop attack the long diagonal. Bobby also threatens mate on the eighth rank [MATE-8] if Boris removes his bishop from guarding his king and Bobby can move his rook to the eighth rank [R8]. Boris will resign [RESIGN] if Bobby threatens mate either on the diagonal or on the eighth rank.

Which of the following will *guarantee* that Boris resigns ?

(A) Bobby sacrifices his knight;
(B) Boris takes the knight with his bishop and exposes KN2;
(C) Boris takes the knight with his bishop and moves the pawn in front of his king;

(D) Boris exposes his KN2 and Bobby can move his Queen to KN7;
(E) Boris takes the knight with his bishop and Bobby can move his rook to the eighth rank;
(F) Boris moves pawn to KN2 and Bobby moves his rook to the eighth rank;
(G) Boris takes the knight with his bishop and either Boris exposes his KN2 or Bobby can move his rook to the eighth rank?

8. *Tennis Court Analogy*

In "The University Case for Preferential Treatment" [1976], Richard Wasserstrom presented an thought provoking analogy. Some have argued that affirmative action programs in medical school admissions are intrinsically unjust. They assume that those who are the best qualified to utilize an academic opportunity or position should get it on those grounds along. This begs the question about what assumptions are built into the loaded term "best qualified", but setting that aside, does justice demand that only the best qualified should get the position or opportunity? [13]

Consider the following case. Suppose there is a town called Wimpleton, which has only one tennis court. Now one day ten tennis players arrive simultaneously to play a game of tennis. Certainly one could not argue that the two best qualified to play should be the ones who get the opportunity to do so. Why not those who desire to play the most? Why not those who will enjoy playing the most? Why not those who will exert the most effort when playing? Similarly, the best qualified to utilize an academic opportunity or position in medical school are not necessarily the ones who should be given those opportunities.

Explicate this argument by analogy, and then discuss whether any of the associated fallacies have been committed.

9. *Freemasons and Tories*

Sherlock Holmes and Dr. Watson were called in by Scotland Yard to investigate the Case of the Missing Crown Jewels. Inspector Lestrade was questioning two suspects—Penwick and Queenie—one of whom was the thief. Each of the two also belonged to one of the two secret societies–the Tories whose members always tell the truth or the Freemasons whose members always tell falsehoods. Holmes overheard the following part of the interrogation:

> "At least one of us is a Freemason," admitted Penwick.
> Queenie replied, "Penny is the thief."
> Holmes casually remarked to Watson, "I have solved the case!"
> "Why Holmes–" Watson replied in amazement.

"How many times must I tell you, Watson, when you have eliminated the impossible, whatever remains—however, improbable—must be the truth!"

Who stole the Crown Jewels? Justify your answer with a truth table analysis for Penwick's assertion.

10. *Courting with Cars*

Four cars are parked in a line in front of the Supreme Court–an AUDI, BMW, BUICK, and CAMARO. Each of the cars is owned by one and only one of the justices—Ruth Bader Ginsberg, Bret Kavanaugh, Sonya Sotomayer, and Clarence Thomas. The cars are one of four colors—*Garnet*, *Kashmir*, *Sable*, and *Tahini*, and no two cars are the same color. You are also given the following clues:

(A) No justice owns a car whose color's name begins with the same letter as its owner's surname.
(B) Neither the BMW nor the BUICK is parked directly behind the first car.
(C) The BMW and the BUICK are owned, respectively, by Thomas and Kavanaugh.
(D) Ginsberg does not own the *Tahini* colored car, nor is the *Tahini* colored car the AUDI.
(E) The *Sable* colored CAMARO is first in line, and the *Kashmir* colored car is last.

What color is the BUICK and in which position is it parked?

	AUDI	BMW	BUICK	CAMARO	Garnet	Kashmir	Sable	Tahini
Ginsberg								
Kavanaugh								
Sotomayor								
Thomas								
1st								
2nd								
3rd								
4th								
Garnet				
Kashmir				
Sable				
Tahini				

11. *Remember Venn?*

Investigate the following syllogisms, and their modifications, using Venn diagrams and the Ladd-Franklin rules for the *reductio* set. The first problem is solved as an example.

No fossil can be crossed in love.
Some oysters can be crossed in love.
∴ No fossils are oysters.

S: fossils
M: things that can be crossed in love
P: oysters

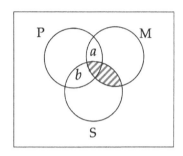

Paraphrase		Reductio Set	
No S are M	≡	*Not*-(Some S is M)	⊘
Some P are M	≡	Some P is M	a
∴ No S are P	≠	Some S is P	b

The syllogism as it stands is *invalid* since the Venn diagram can represent its *reductio* set without contradiction. Furthermore, the first Ladd-Franklin rule is violated. (When the first rule is violated, the second rule can't be applied, but failing any of the three conditions results in a invalid syllogism.)

Which of the following replacements in the syllogism above results in a valid syllogism?

I. Replacing premise 1 with "Only fossils can be crossed in love."
II. Replacing premise 2 with "Some things that can be crossed in love are oysters."
III. Replacing the conclusion with "Some fossils are oysters."

a. I only.
b. II only.
c. III only.
d. Both I and III together.
e. Either II or III, but not both.

You can use the Ladd-Franklin rules *forwards* (to detect *invalidity*) and *backwards* (to redirect into *validity*), which can save a lot of time on a standardized test involving syllogisms. Notice that Option II replaces "Some P are M with "Some M are P", which are logically equivalent. Answers (ii) and (v) can be eliminated right off the bat. Answer (i) is also incorrect since Option I alone does not repair the syllogism to satisfy the first Ladd-Franklin rule. Answer (iii) is incorrect because it satisfies the first Ladd-Franklin rule but not the third.

Let's verify that answer (iv) is correct. In terms of the Ladd-Franklin rules, there are two structural problems with the *reductio* set:

There are *two* positive existentials, *not one*, as required by LF$_1$.

The pair of negative existentials must have a pair of *contradictory* terms, as required by LF$_2$.

The first problem can be fixed by replacing the conclusion with "*Some fossils are oysters*," which when turned into its opposite or negation for the *reductio* set, provides the second negative existential. This satisfies the first Ladd-Franklin rule that there is exactly one positive existential and makes the second Ladd-Franklin rules applicable: the term common to the pair of negative existentials (namely, S) must be a contradictory pair. This condition fails but can be satisfied in several ways.

One way is to replace the first negative existential in the *reductio* set with "*Not-(Some M is not-S)*". Working backward, this means that the first premise can be replaced by "All M is S" or "*All things crossed in love are fossils*". However, "All M is S" is equivalent to "Only S is M" or "*Only fossils are crossed in love.*" Combining Options I and III fixes both structural problems: the correct answer is (iv).

Replacement Options		Reductio Set	
I. Only S is M	≡	*Not*-(Some M is *not*-S)	⊘
II. Some P are M	≡	Some M is P	*a*
III. ∴ Some P is S	≠	*Not*-(Some P is S)	⊘

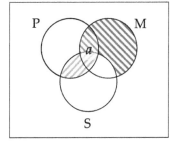

The Venn diagram for the revised *reductio* set confirms our analysis. Shading out the two negative existentials first, we are forced to place a letter in a shaded out area. This is a *forced* contradiction when trying to represent the *reductio* set. It is therefore *impossible* for the premises of the revised argument to be true while its conclusion is false. The revised syllogism is valid.

Here are some additional problems.

(A) Some bows are not made of angel hair.
No ice cream castles in the air are made of angel hair.
Therefore, some bows are ice cream castles in the air.

Which replacements result in a valid syllogism?

I. Replacing premise 1 with "Nothing that isn't made of angel hair is a bow or flow.
II. Replacing premise 2 with "All ice cream castles in the air are made of angel hair."
III. Replacing the conclusion with "Some bows aren't ice cream castles in the air."

a. I only
b. II only
c. III only
d. Either I or II, but not together
e. Both II and III together.

(B) Some cats have short hair.
Nothing with long whiskers has short hair.
Therefore, some cats have long whiskers.

Which of the following modifications results in a valid syllogism?

I. Replacing premise 2 with "All who don't have long whiskers don't have short hair."
II. Replacing premise 2 with "Nothing with short hair has long whiskers."
III. Replacing the conclusion with "Some cats do not have long whiskers."

a. I only.
b. II only.
c. III only.
d. Either I or II.
e. Either I or III.

(C) All inhabitants of Middle Earth love creature comforts.
Some Hobbits love creature comforts.
∴ Some Hobbits are inhabitants of Middle Earth.

Which of the following replacements result in a valid syllogism?

I. Replacing premise 1 with "No inhabitants of Middle Earth are lovers of creature comforts."
II. Replacing premise 2 with "Some who love creature comforts are Hobbits."
III. Replacing the conclusion with "Some Hobbits are not inhabitants of Middle Earth."

a. I only.
b. II only.
c. III only.
d. Both I and II only.
e. Both I and III only.

12. *Doubles or Nothing*

A game consists of rolling a pair of dice–one die is red and the other green. On the red die you know you've rolled an odd number, but you don't know which. You have no information about which of the six numbers you rolled on your green die. Winning depends on the sum of the numbers showing on the pair of dice. A doubles is thrown when both dice show the same number. Given this information, which is the better bet?

(A) You win if the sum is greater than or equal to seven but is not a doubles.
(B) You win if the sum is less than seven or is a doubles.

Group III

13. *A Murderous Analogy*

Explicate the following using the argument schema for analogies:

"My client, Ms. Jones, put one live cartridge into a six-chambered revolver, spun the chamber, aimed the gun at Smith and pulled the trigger hoping to kill Smith. Smith was shot. We submit nevertheless that we cannot say that Ms. Jones killed Smith intentionally. Suppose that Brown in an ordinary game of dice hopes to throw a six and does do so, we do not say that Brown threw the six intentionally."

Then evaluate the analogy using the three associated fallacies.

14. *Truth-Teller Triangles*

Knights always tell the truth and *Knaves* always tell falsehoods. Individuals are of the *same type* if they're both Knights or both Knaves. In the following scenarios, Art, Gwen, and Lance are each either a Knight or a Knave. Try to determine the status of each whenever possible. In one of the scenarios the status of only one of the three can be determined.

(A) Art says, "Gwen is a Knave."
    Gwen says, "Art and Lance are of the same type."
(B) Art says, "All of us are Knaves."
    Gwen says, "Exactly one of us is a Knight."
(C) Art says, "All of us are Knaves."
    Gwen says, "Exactly one of us is a Knave."

15. *Positively Logical*

This time you are traveling with Alfred, Rudolf, and Kurt. You know that exactly one of your traveling companions is a Logical Positivist.

Alfred says, "Kurt is a Logical Positivist."
Rudolf says, "I am not a Logical Positivist."
Kurt says, "At least two of us are Knaves."

Suppose you need to choose one of the three as your traveling companion, and it is more important to you that this traveling companion not be a Logical Positivist than not be a Knave. Whom should you choose?

16. Pyramid Probabilities

A game is played with a pair of tetrahedral dice. (A *tetrahedron* is a pyramid each of whose four sides is an equilateral triangle.) The faces of each die are numbered 1, 2, 3, and 4. Assuming that the dice are fair, and given that the numbers you have tossed on the pair of dice are *different*, which is the best bet?

(A) the sum is either a two or a three
(B) the sum is either a four or a five
(C) the sum is even

17. Preppies

Four Preppies nicknamed Muffy, Buff, Skip, and Kip each drive a car given to them by wealthy relatives. These cars are the *Volaré*, the *Marquis*, the *Impala*, and the *LTD*. No two Preppies drive the same type of car. These Preppies, are all dressed in appropriate Preppy attire–a lime-green golf shirt, a bubblegum-pink sweater, penny loafers, and alligator underwear. Each of the Preppies wears one of these items, and no two Preppies wear the (same type) of item. You also know the following facts.

(A) Skip's Auntie bought him his *Marquis* on the condition that he promise not to put any bumper stickers on it.
(B) Neither the *Volaré* nor the *Marquis* would go with Muffy's bubblegum-pink sweater, and she would die before she would drive a car that clashed with her clothes.
(C) The Preppy who drives the *LTD* wears a lime-green golf shirt.
(D) The Preppies with the lime-green golf shirt and the alligator underwear would not exhibit a *Coors* decal on their bumpers (they would never exhibit such a decal on their cars either).
(E) Neither Buff nor Kip would wear penny loafers, and Buff won't even wear a lime-green golf shirt.

Skip always keeps his promises. Which Preppy has the *Coors* bumper sticker (on their car)?

18. Blind Dates

On a TV game show, there are three doors on stage, behind each of which is a prize. On the day that you are a contestant, behind two of the doors is a crate of dates, and behind the third door is a $1,000 bill. You are to choose one door at random, and you will receive what is behind that door. Your chances are 1/3 that you will get the desired $1,000.

(A) You choose one of the doors. The game show host opens one of the other doors to reveal a box of dates. Have your chances of winning now increased to 1/2?

(B) After revealing that one of the other doors has a box of dates, the game show host gives you a chance to change your choice. Should you switch or keep to your first choice?

19. *Brave New Worlds*

In a certain society of people there are three marriage groups–Alphas, Betas, and Deltas. A marriage between a male and female is allowed if and only if they belong to the same group. The sons of Alphas are Deltas, but the daughters of Alphas are Betas. The sons of Betas are Alphas, but the daughters of Betas are Deltas. The sons of Deltas are Betas, but the daughters of Deltas are Alphas.

(A) Which of the following is a complete and accurate list of the groups to which the grandchildren of an Alpha could possibly belong?
   a. Alphas
   b. Alphas, Betas
   c. Betas, Deltas
   d. Alphas, Deltas
   e. Alphas, Betas, Deltas

(B) Under the rules given, may a male marry his niece?
   a. No.
   b. Yes, if she is his brother's daughter.
   c. Yes, if she is his sister's daughter.
   d. Yes, if she is his wife's brother's daughter.
   e. Yes, if she is his wife's sister's daughter.

(C) Which of the following kinds of aunts may a male marry?
   a. only his mother's sister
   b. only his father's sister
   c. only the widow of his father's brother
   d. only his mother's sister or his father's sister
   e. his mother's sister, his father's sister, or the widow of his father's brother

(D) Which of the following statements can be inferred from the rules given?
   I. A brother and sister may not marry.
   II. A female may not marry her father.
   III. A male may not marry his granddaughter.

   a. I only
   b. II only
   c. III only
   d. Both I and II only
   e. All of I, II, and III

20. *Rotten Eggs*

An egg carton is filled with a dozen eggs. Each of the twelve compartments in the egg carton contains one and only one egg. The compartments are labeled with the letters A through L.

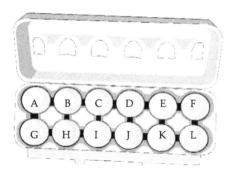

Six of the eggs are rotten and six of the eggs are fresh. The eggs in compartments H, I, J, and K are all fresh. The eggs in compartments A, B, C, D, E, and F are not all rotten, but the eggs in compartments A and F are rotten. Eggs in compartments that are immediately adjacent to each other either horizontally, vertically, or diagonally are called *neighbors*. The eggs in compartments H and K each have exactly three rotten neighbors.

(A) Which of the following must be true?

   I. Either C or D has a rotten egg.
   II. Either G or L has a rotten egg.
   III. Either B or E has a rotten egg.

   a. I only
   b. II only
   c. III only
   d. I and II only
   e. None of the above

(B) If compartments B and G contain rotten eggs, then which of the following must be the case?

   a. C does not have a rotten egg.
   b. D does not have a rotten egg.
   c. E does not have a rotten egg.
   d. D has a rotten egg.
   e. E has a rotten egg.

(C) Assume the egg in compartment I has exactly one rotten egg as a neighbor. If the egg in compartment J has exactly two rotten eggs as neighbors, which of the following must contain eggs that are not rotten?

   a. C
   b. D
   c. Both C and D
   d. Either G or L but not both
   e. None of the above

(D) Assuming that the egg in compartment I has exactly one rotten neighbor and that the egg in compartment J has exactly one rotten neighbor, which of the following must be true?

   a. Both B and D must have rotten eggs.
   b. Both C and E must have rotten eggs.
   c. Both C and D must have rotten eggs.
   d. Either B or D has a rotten egg.
   e. None of the above.

(E) Which of the following is impossible given that the eggs in compartments I and J each have two rotten eggs as neighbors?

   I. G and L have rotten eggs.
   II. B and E have rotten eggs.
   III. C and D have fresh eggs.

   a. I only
   b. II only
   c. III only
   d. Both I and II only
   e. Both II and III only

21. *Sorting Sorites with Lewis Carroll Diagrams*

Sorites are chains of syllogisms in which the intermediate conclusions have been omitted. What conclusion can be drawn from the following sorites composed by Lewis Carroll?

(A) Babies are illogical.
Nobody is despised who can manage a crocodile.
Illogical persons are despised.
∴ _____

From the first and third premises deduce the intermediate conclusion:

∴ Babies are despised.

Combining this conclusion with the second premise, allows us to deduce the final conclusion:

∴ Babies cannot manage a crocodile.

Although it is impossible to draw a Venn diagram for four terms using only circles of the same size (e.g., ovals are required), Lewis Carroll invented a diagram with four terms that uses overlapping rectangles. The large red (*horizontal*) and to blue (*vertical*) rectangles intersect in the purple square creating 3 areas. The smaller yellow (*horizontal*) and green rectangles (*vertical*) create 12 more areas, which in addition to the white area represents all 16 possibilities.

Solving the four term sorites using the Lewis Carroll diagram is completely analogous to using Venn diagrams. First, we construct the *reductio* set as before. As before, we "shade out" before "lettering in".

All B are I	≡	*Not*-(Some B is *not*-I)	
No D are M	≡	*Not*-(Some D is M)	
All I are D	≡	*Not*-(Some I is *not*-D)	
∴ No B are M	≠	Some B is M	a

Carroll invented a "Game of Logic" with rectangular areas equivalent to Venn diagrams that can be played with black and white counters (e.g., like Go stones) which when placed in a given partition corresponds to "shading out" and "lettering in" a given logical possibility. Counters can be also be used on the four term Lewis Carroll diagram as well.

The validity of our conclusion is represented by the fact, we're forced to place a letter in a shaded out area.

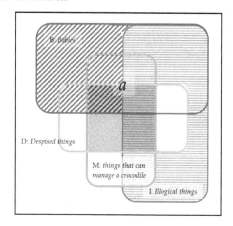

Fill in the conclusions to the sorites composed by Lewis Carroll:

(B) My saucepans are the only things I have that are made of tin;
I find all your presents very useful;
None of my saucepans are of the slightest use.
∴ _____

(C) No potatoes of mine, that are new, have been boiled;
All my potatoes in this dish are fit to eat;
No unboiled potatoes of mine are fit to eat.
∴ _____

(D) Everyone who is sane can do Logic;
No lunatics are fit to serve on a jury;
None of your sons can do Logic.
∴ _____

(E) There are no pencils of mine in this box;
No sugar-plums of mine are cigars;
The whole of my property, that is not in this box, consists of cigars.
∴ _____

22. Alan and the Enigmatics

On planet Enigma, the inhabitants (called *Enigmatics*) are either Humans or Androids but not both. The Androids and Humans are indistinguishable by appearance alone. Half the Humans always tell the truth and the other half always tell lies. Similarly, half the Androids always tell the truth and the other half always tell lies.

Alan arrives on Enigma and overhears the statements made by Enigmatics (left-hand column) and he immediately makes a correspondingly *correct* deduction about the speaker (right-hand column). Match each Enigmatic statement with Alan's deduction.

Enigmatic Statements			Alan Deduces the Speaker is
I'm a truthful Android	(A)	(1)	a truthful Human
I'm a lying Human	(B)	(2)	a lying Android.
I'm not a truthful Android	(C)	(3)	an Android (but whether truthful or lying, Alan can't tell.
I'm either a truthful Android or a lying Human	(D)	(4)	an Android if the statement is *true* (but if *false*, Alan can't tell).
I'm an Android	(E)	(5)	a truthful Android or a lying Human (but Alan can't tell which).

5
Computational Magic

Mathematical reasoning may be regarded rather schematically as the exercise of a combination of two facilities, which we may call intuition and ingenuity. The activity of the intuition consists in making spontaneous judgements which are not the result of conscious trains of reasoning... The exercise of ingenuity in mathematics consists in aiding the intuition through suitable arrangements of propositions, and perhaps geometrical figures or drawings.[1]

—ALAN TURING (1912-1954)

Computational thinking is a set of problem-solving skills that arose from thinking about the nature of computation. Included in this skill set are constructing and "debugging" algorithms, designing computer models or simulations, and reverse engineering algorithms to figure how they work or to discover their weaknesses. Computational thinking also encompasses design issues. The interface between computers and humans is critical to their effectiveness. What is the best *data architecture* for storing, updating, and recalling information? How can you design the *interface* so that using the algorithm is intuitive and minimizes human errors? In this section we give a preview of some of these computational problem solving skills. Computational thinking will be as important in the 21st century as reading, writing and arithmetic were in earlier centuries.

A "Digital Computer"

What is an algorithm? An *algorithm* is a series of step-by-step instructions for obtaining a result such that each step is clearly defined, executable, and requires no ingenuity. Let's illustrate the idea of an *algorithm* by describing how you can use your hands to multiply numbers from 6 – 10. Here we shall assume you (the computer) have memorized (or stored in a "look up" table in memory) the multiplication tables for the numbers 1 – 5 and that you can add.

Our "digital computer" will have ten locations with addresses, which we call stores or memory locations. The numbers 6, 7, 8, 9, 10 are stored in these locations. You can write the numerals on the fingertips of each hand, beginning with the little fingers and ending with the thumbs. You are the central processing unit which executes the following algorithm or set of step-by-step instructions.

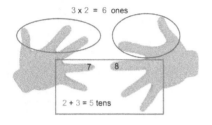

$7 \times 8 = 50 + 6 = 56$

1. To multiply two numbers from 6 to 10, simply touch the fingers which store them. The diagram shows to multiply 7×8.
2. *Add* the number of fingers on each hand that are touching and below. This is the *number of tens*. In this example, we have $2 + 3 = 5$ tens or 50.
3. *Multiply* the number of fingers above those touching on each hand. This is the *number of ones*. In this example, we have $3 \times 2 = 6$ ones or 6.
4. Then the product of the original two numbers is the sum of the *number of tens* and the *number of ones*. In this example, $7 \times 8 = 50 + 6 = 56$.

Practice this algorithm a few more times. The crux of the algorithm is easy to remember:

- *Add* the number of fingers touching below to compute the tens.
- *Multiply* the number of fingers above to compute the ones.

Verify the algorithm works for all products of the numbers 6 - 10.

A few comments are in order. First, following instructions doesn't require any ingenuity. These "mechanical" instructions could be carried out by a machine. Secondly, the algorithm will terminate in a finite number of steps and is guaranteed to give the correct answer. Thirdly, you don't have to understand *why* the algorithm works in order to use it. Later we'll explain why the algorithm works, both algebraically and geometrically. For now, the key point is that mechanically following, or instantiating, an algorithm doesn't require that you understand *how* the algorithm works.

One entertaining way to explore algorithms is through self-working card tricks. Automatic card tricks are algorithms. By manipulating a deck of cards, you can acquire some "hands-on" experience of following an algorithm and exploring some key elements of computational problem solving: reverse engineering, designing data architecture, and creating an interface adapted to human experience.

Dialing Up An Algorithm

The *Cell Phone Trick* was reverse-engineered from a card trick I learned from a young magician that I met at a farmers market while driving through Oregon. (You can see the trick performed, which goes by the name "1 out of 357,689,142" on *YouTube*). For ease of exposition, the Magician will be alternatively be referred to as "he" or "she", and the volunteer will be of the opposite sex.

The Effect: A Magician takes a deck of cards Ace, 2, 3, 4, 5, 6, 7, 8, 9. The Magician hands the spectator a folded piece of paper to put in her pocket.

After explaining to the spectator how she can randomize the cards by calling out "Down" or "Switch," the Magician rearranges the deck several times according to the spectator's choices. When the spectator decides that the deck has been thoroughly rearranged, the Magician stops and deals out the cards. The Magician instructs the spectator to open up the piece of paper in her pocket. The paper correctly predicts the arrangement of the cards.

When I started thinking about how to make the trick more fun and versatile, I got the idea of using 10 cards to spell out a cell phone number—either the spectator's own or a pre-arranged cell phone hidden in the vicinity that would ring when the spectator is asked to dial the number.

Let's see if we can reverse engineer the original trick to create a new effect. We'll use the Ace for 1 and the 10 for the digit 0. For now, you'll be playing the role of both the Magician and the Spectator.

The Cell Phone Set-Up: Begin with a deck arranged face down from top to bottom as follows: Ace, 2, 3, 4, 5, 6, 7, 8, 9, 10. We will describe an algorithm that sets up the trick.

1. Deal the top card, the Ace, face-up on the table.
2. Next, deal the second card, the deuce, face-up on top of the Ace and say "Down."
3. Flip over next *pair* of cards, the 3 and the 4, and let the spectator look at them face up as you switch their order and say "Switch." Notice that you've switched the order of the 3 and 4 as you place them on top of the face up pile.
4. Repeat steps 2 and 3, until you've gone through the entire deck of 10 cards.
5. Next take the face up pile of cards and turn them deck face down.
6. Deal the top card down on the table face up, saying "Down." Then flip over the next pair of cards face-up, saying "Switch," as you switch their order before placing them on the face up deck. These are you two choices–"Down" and "Switch."
7. Turn the three face-up cards face-down and drop the remaining deck on top of them (face down).

The Trick:

8. Holding the cards *facing down*, go through the entire deck —keeping the cards face down—letting the volunteer instruct you at each step to either deal the card Down or to Switch a pair of cards.
9. Repeat step 1 an *even* number of times.

The Reveal:

10. Deal out the cards one at a time.
11. The order of the cards is: 3 5 7 - 6 8 10 – 9 1 4 2.

This trick cannot be repeated too many times without giving away the secret. So, let's reverse engineer the trick. Instead of having the same outcome every time, let's vary the trick to produce a pre-determined cell phone number. We'll use a spreadsheet to create a computer model of the card trick. The architecture of the data base is simple. There will be three columns:

	A	B	C
1	1	3	8
2	2	5	10
3	3	7	1
4	4	6	9
5	5	8	2
6	6	0	4
7	7	9	3
8	8	1	5
9	9	4	7
10	10	2	6
	INPUT	KEY	OUTPUT

Column A: The INPUT will be a 10-digit cell phone number.
Column B: The KEY CODE in this case is 3 5 7 - 6 8 10 - 9 1 4 2.
Column C: The OUTPUT will be the set-up for the phone trick.

Since the Ace or 1 ended up in the 8th position of the keycode, we want the 1st position in the set-up to be the 8th number in the input. Hence, in cell C1 we enter "=A8." Since, the 2 ended up in the 10th position of the keycode, so in cell C2, we enter "= A10... ." The 10th number ended up in the 6th position and so in cell C10, we enter "= A6." This spreadsheet now calculates for a given input, the set-up. It is also flexible insofar as we can construct different algorithms with correspondingly different key codes.

It is a good idea to verify that you've designed the data base correctly. If you input 1234567890, you should get 8019243576. What do you get for the input 0S6-AAN-1FL0? To test your understanding of reverse engineering, what's the input for obtaining as the output PUZZLINGLY (a 34-point Scrabble word)?

Points to Ponder

- You can go through the deck calling out "DOWN" or "SWITCH" an *even* or *odd* number of times. What difference does this make?
- It is important to "sell the trick." When the cards are face-up go slow enough to that you, or later the volunteer, can see that the order of the cards is actually *reversed* when you "switch" the cards.

So far, we've learned how to follow an algorithm and how to create a data architecture using a spreadsheet that models the original card trick. Next we'll take into consideration the human element of design, an unduly neglected part of computational problem solving.

Programming for People

How many times have you had the frustrating experience of filling in an on-line form that required you to re-enter data, over and over, because the designers didn't take into account likely human errors–e.g. whether it asked for your first name first or you last name first, whether it required you to put only the last two digits of your birth year, or to include dashes between parts of your phone number. Sometimes your entire page of input is erased when you go back to correct a single mistake.

Suppose you want to be able to compute the set-up for the Cell Phone Algorithm "on the fly" with a deck of cards without relying on a spreadsheet? How can you design a human interface which accomplishes what the spreadsheet does utilizing facts about human experience to minimize human input and calculation errors? One feature of daily life is dialing a cell phone number.

Suppose that you are performing the trick for a volunteer hands you a series of 10 cards face down representing his cell phone number. How can you, without looking at the cards, quickly arrange the cards for the set-up? Given a deck with a secret cell number:

1. Deal the cards containing the secret phone number onto an imaginary cell phone pad as if you're dialing the Key Code: 375-680-9142.
2. Pick up the cards in numerical order, i.e., positions 1, 2, 3, 4, 5, 6, 7, 8, 9, 0.

The set-up algorithm now as easy as dialing up a frequently used phone number. Thinking about designing the human interface so that user's experience is simple and natural as possible is an important, often overlooked part, of computational problem solving.

Have you ever wondered why computer programmers code numbers in binary? And why do they like to start counting with 0 instead of 1? We will see why in a moment.

31 Flavours of Freedom

Once upon a time, there lived a very logical Indian Princess, who returned to India from America to attend the wedding of her older sister in the great Falaknuma Palace. Her father decided that she too should get married very soon, but knew his daughter would not simply give in to his wishes, so he decided to play a game. The King said he would pick just one of the 31 flavours of ice cream offered at the wedding banquet. If she could correctly guess his chosen flavour, she would have 31 more months of freedom.

"What guarantee do I have you won't change your choice after I choose?" she asked.

"Well, I'll write it down on a piece of paper," he replied. Now the princess was logical and she proposed a different plan, "I will assign a list of flavours to each of the five groomsmen, each of whom was wearing a different colored cumberbund. All you will do is to answer 'yes' or 'no' if your chosen ice cream is on their lists. Then I will make my choice."

The king thought to himself, "there would be six flavours on each list, with one left over. So, the chances of her guessing correctly is still about one out of six." He countered, "if on each of your lists there are at least 12 flavours, then I'll agree."

She replied, "I'll put more than a dozen flavours on each list. Moreover, if I can choose correctly three times in a row, then you will agree that I will decide who and when and if I shall marry." The king was taken aback by his daughter's confidence and brashness, but he agreed to her terms.

When the trial was over, the Princess had chosen the correct flavour three times in a row, answering within seconds of knowing the answers from the groomsmen. How did she accomplish this feat?

The following is the chart the princess used for coding and decoding the 31 flavours. The binary code for flavour 29, sweet potato maple walnut, is 11101, because 29 = 16 + 8 + 4 + 1. Reversing the procedure, the flavour for the binary code is flavour 13 = 8 + 4 + 1, which is Huckleberry.

What are the binary codes for flavours 3, 13, and 31?

What flavours correspond to codes 11101, 01100, 01101?

There are 10 types of people in the world, those who understand binary, and those who don't.

The 31 Flavours

Can you think of a way to use your hand as a digital binary computer?

Most of us learned how to count using our fingers in base 10. In the standard base 10 system, the values of the columns are powers of 10. In binary, or base 2, the values of the columns are power of 2 :
$2^0 = 1, \ 2^1 = 2,$
$2^2 = 4, \ 2^3 = 8,$
$2^4 = 16, \ 2^5 = 32,$ etc.

		16	8	4	2	1
		0	0	0	0	0
Almond	1	0	0	0	0	1
Bastani Sonnati (Persian)	2	0	0	0	1	0
Butter Pecan	3	0	0	0	1	1
Cayenne Chocolate	4	0	0	1	0	0
Chocolate Chip Cookie Dough	5	0	0	1	0	1
Coconut Almond Chip	6	0	0	1	1	0
Dulce de Leche	7	0	0	1	1	1
English Toffee	8	0	1	0	0	0
Goat Cheese Beet Swirl	9	0	1	0	0	1
Green Tea	10	0	1	0	1	0
Honey Avocado	11	0	1	0	1	1
Honeyjack & Coke	12	0	1	1	0	0
Huckleberry	13	0	1	1	0	1
Jalapeño	14	0	1	1	1	0
Lavender Honey	15	0	1	1	1	1
Les Bourgeois & Ghirardelli	16	1	0	0	0	0
Madagascar Vanilla	17	1	0	0	0	1
Mango	18	1	0	0	1	0
Mint Chocolate Chip	19	1	0	0	1	1
Moose Tracks	20	1	0	1	0	0
Passion Fruit	21	1	0	1	0	1
Pistachio	22	1	0	1	1	0
Peanut Butter	23	1	0	1	1	1
Pralines & Cream	24	1	1	0	0	0
Red Velvet	25	1	1	0	0	1
Rum Raisin	26	1	1	0	1	0
Spumoni	27	1	1	0	1	1
Strawberry	28	1	1	1	0	0
Sweet Potato Maple Walnut	29	1	1	1	0	1
Ube (Philippines, purple yam)	30	1	1	1	1	0
Vietnamese Coffee	31	1	1	1	1	1

The Josephus Problem

The first century historian Titus Flavius Josephus (37 – c. 100) in *The Jewish War* describes a deadly dilemma. At the siege of Yodfat, Roman soldiers trapped Josephus and 40 other soldiers in a cave. The Jewish soldiers decided self-sacrifice was more honorable than being captured alive. To minimize the need for suicide, they devised a plan. Arranging themselves in a circle facing each other:

1. The first soldier chosen at random would slay the solider to his left.
2. Continuing clockwise, the next soldier slays the soldier to his left, etc.
3. This process was to be continued until there was one man left, who would then commit suicide.

Counting clockwise from the first soldier, what's the position of the last man standing?

Here's an algorithm to determine the last man standing, where n is the number of soldiers:

1. Convert n into a binary string.
2. Move the leftmost 1 of the binary string to the right end of the string.
3. Convert the new string back into a number.

If you begin counting the first soldier as 1 and count clockwise, the algorithm gives you the position of the last soldier standing.

Let's model the Josephus problem using a deck of cards. Take deck of 41 cards. With the cards facing you, turn the 18th (not the 19th) card from the top the other way around. Now you're going to perform a "down-and-under" shuffle:

1. Discard the first card.
2. Move the next card to the bottom of the deck.
3. Repeat steps 1 and 2 until there's one card left.

It should be the chosen card.

Here's an algorithm for determining the position of the last using the "down-and-under" shuffle card for a deck of n cards:

1. From n subtract the highest power of 2 less than n.
2. Take that remainder and double it.
3. This number is the position of the last card.

For example, if $n = 41$ then the highest power of 2 less than 41 is $2^5 = 32$. Subtracting 32 from 41, we have a remainder of 9, which doubled is 18. The last card standing will be the 18th card from the top of the deck. The "down-and-under" shuffle is a "hands-on" method for testing algorithms.

Mystery #1: Why do the two algorithms differ by one?

The answer to this mystery is simple. The counting in the original Josephus problem began with the first *survivor*, but the "down-and-under" shuffle began counting with the first soldier *slain*. To make the two operations identical, simply move one card from the bottom of the deck to the top and then use a "*bottom-and-down*" shuffle. The binary algorithm often begin counting at zero, so you need to remember to add, or subtract, 1.

Mystery #2: Why do computer programmers, and logicians, like to start counting with 0 instead of 1?

This reminds me of a story. When I took classes from the great 20th century logician Alonzo Church, he would often pass out handouts. Standing in front of row of 7 students, Professor Church would count "0, 1, 2, 3, 4, 5, 6," and then hand the pile of papers to the first student. Eventually it occurred to me that this may have been due to the fact that, according to von Neumann's logical construction of the natural numbers, the number $7 = \{0, 1, 2, 3, 4, 5, 6\}$! More likely Professor Church was expressing his perversely logical sense of humor. Most of us students simply listened to him in unquestioning silence, too awed to ask.

John Horton Conway (1937-2020) in *The Book of Numbers* tells another story (probably apocryphal) about the Polish mathematician Waclaw Sierpinski (1882 - 1969). Sierpinski was once worried that he had lost one piece of luggage. "No dear," his wife said, "all six pieces are here!" Sierpinski replied, "That can't be true. I've counted them several times: zero, one, two, three, four, five." Counting from 0 has its own problems: what would Sierpinski's count be if all six pieces of luggage were missing?[2]

The art of doing mathematics consists in finding the special case that contains all the germs of generality.
—Quote attributed to DAVID HILBERT[4]

When faced with a problem you don't understand, it is often useful to find a simpler problem you don't understand. The simpler problem should be complex enough to exhibit the most important features of the original problem, yet simple enough to comprehend as a whole.

Imagine 12 soldiers positioned at the hours of a circular clock. Suppose that the soldier standing at 12:00 or noon was chosen as the first person. The diagram below shows the positions or hours of the clock with their binary codes.

- The red circles represent the soldiers eliminated in the first round. These are the ones with 1 in the 1's column of their binary code.
- The green circles represent the soldiers eliminated in the second round. These are the soldiers with 1 in the 2nd (or 2's) column in their binary code.
- The orange circles represent the soldiers eliminated in the third round. These are the solider with 1 is the 3rd (or 4's) column in their binary code.
- The blue circle at 8 o'clock is the last man standing. This is the soldier who has a 1 in the 4th (or 8's) column of his binary code and who hasn't previously been eliminated.

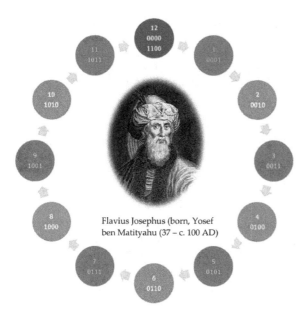

Flavius Josephus (born, Yosef ben Matityahu (37 – c. 100 AD)

The string algorithm begins counting at zero. It tells us that the 9th position moving clockwise will contain the last man standing. Since the counting starts at zero (or noon), the 9th position is 8 o'clock. The algorithm for the "down-and-bottom" shuffle begins counting at 1 o'clock with the first soldier to be eliminated. Since the counting starts at 1 o'clock, counting to 8 also brings you to 8 o'clock.

Let's begin with a deck of 12 face-up cards in numerical order expressed in binary. In round 1 all the cards whose code has a 1 in the 1's column are eliminated; in round 2 all the cards whose code has a 1 in the 2's column are eliminated; in round 3 all the cards whose code has a 1 in the 4's column are eliminated. This leaves the 8 as the last card standing, the only card remaining with a 1 in the 8's column. Binary code is a simple way of seeing and saying what's going on.

Q♥ 1100	J♦ 1011	10♠ 1010	9♦ 1001	8♣ 1000	7♦ 0111	6♠ 0110	5♦ 0101	4♥ 0100	3♦ 0011	2♠ 0010	A♦ 0001
	J♦ 1011		9♦ 1001		7♦ 0111		5♦ 0101		3♦ 0011		A♦ 0001
		10♠ 1010				6♠ 0110				2♠ 0010	
Q♥ 1100								4♥ 0100			
				8♣ 1000							

Mystery #3: Why do the two algorithms amount to the same thing?

The "down-and-under" algorithm calculates the last card as follows:

1. Subtract the largest power of 2 that you can from n.
2. Double the remainder.

The largest power of 2 less than 12 is $2^3 = 8$, and so $12 - 8 = 4$ and $4 \times 2 = 8$. The last soldier standing is at 8 o'clock.

The binary string algorithm also has two steps. Converting the number of soldiers, $n = 12$, into a binary string yields 1100. The string algorithm instructs you to:

1. +100   delete the leftmost 1.
2.   1001   add a 1 to the end of the string.

Deleting the leftmost 1 in a binary string is the same as subtracting the largest power of 2. Moving the 1 to the right end of the string does two things at once: (1) it shifts the remaining string to the left, which is the same as *doubling* it; and (2) it increases the count by +1 (which is necessary if you start counting at zero). Thus, the two algorithms are equivalent. The only difference is where they start counting. The "down-and-under" algorithm starts counting from the first soldier *eliminated*. The string algorithm starts counting with the first soldier to do the *eliminating*.

We leave a final mystery for the reader: give a clear and concise explanation why the algorithms work!

The Josephus story ends with a sobering concession to the power of self-preservation. You might be wondering, how did Josephus survive to write this history? The last man standing would have to face the moral dilemma of choosing between the lesser of two evils–committing suicide or breaking a promise. Apparently, the last *two* standing did not follow the algorithm to its bitter end, only to its *binary* end. Josephus explained: "by luck or possibly by the hand of God" he and another man remained until the end and surrendered to the Romans rather than killing themselves.

The Digital Computer Explained

Let's explain the digital multiplication algorithm. One point should already be clear: merely *following* an algorithm is different from *understanding* what the algorithm is doing. We give two different proofs about why finger multiplication works. The first is algebraic and the second is geometric.

For an algebraic proof, we simply express the algorithm abstractly using variables and then verify that the algebra works out. Suppose we're multiplying 7×8, with 7 on the left and 8 on the right. The number of fingers touching or below is represented by $7 - 5 = 2$ on the left and $8 - 5 = 3$. The sum of these two is the number of 10's. The number above those touching are $10 - 7 = 3$ and $10 - 8 = 2$. The product of these two numbers is the number of 1's. Replacing 7 and 8 with variables verify the following equation:

$$X \times Y = [(X - 5) + (Y - 5)] \, 10 + [(10 - X) \times (10 - Y)]$$

Now for a geometric proof. Imagine a 10 × 10 chessboard. The area of the entire board 100, and the white rectangular represents 7 × 8.

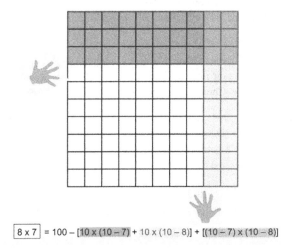

8 x 7 = 100 − [10 x (10 − 7) + 10 x (10 − 8)] + [(10 − 7) x (10 − 8)]

We can calculate the area of white rectangle by subtracting the red and yellow rectangles from 100 and then adding back their overlap, the orange rectangle, which was subtracted twice.

Label the vertical axis X and the horizontal axis Y (although this reverses convention, it facilitates comprehension). It is easy to verify that the following equation is valid:

$$X \times Y = 100 - [10 - X) + (10 - Y)] 10 + [(10 - X) \times (10 - Y)]$$

A little more algebraic manipulation shows our two equations are indeed equivalent.

Alan Turing: the Enigma

Computational thinking is a set of problem-solving skills that arose from thinking about the nature of computation.

Kurt Gödel   Alonzo Church   Alan Turing

Kurt Gödel (1906-1978) the greatest logician of the 20th century known for his Incompleteness Theorems; one of the few members of Princeton's Institute for Advanced Studies (IAS) who walked as an equal with Einstein, although born the year *after* Einstein's *mirabilis annus* of 1905.

Alonzo Church (1903-1995) one of the great logicians of the 20th century best known for the lambda calculus, the Church–Turing thesis, Church's Theorem proving the undecidability of the *Entscheidungsproblem*, and his unfinished Frege–Church Language of Sense and Denotation.

Alan Turing (1912-1954) the greatest computer scientist of the 20th century before computer science was a science; his seminal ideas on computation, universal Turing Machines, the Turing Test, artificial intelligence (AI), and cryptography formed the foundations of that new science.

The most famous, and persuasive, analysis of computation was set forth by the British mathematician Alan Turing in a pioneering paper "On Computable Numbers, with an Application to the *Entscheidungsproblem*" [1936].

In 2019 *The New York Times* published a belated obituary for Turing (in a series of obituaries about remarkable people whose deaths went unreported in the *NYT*), entitled "Overlooked No More: Alan Turing, Condemned Code Breaker and Computer Visionary." Turing solved the famous *Entscheidungsproblem* or "Decision Problem" posed by the German mathematician David Hilbert (1862-1943) challenging mathematicians to find a "decision procedure," or what today is known as an algorithm, to solve any mathematical problem.

Turing realized there had to be real numbers that cannot be computed: there are *countably* many computer programs but *uncountably* many real numbers. To prove this result as a mathematical theorem, Turing had to provide a rigorous mathematical analysis of a computable procedure. Using his definition and the kind of reasoning deployed in Gödel's Incompleteness Theorem, Turing proved the *Unsolvability of the Halting Problem*: given an arbitrary algorithm, there is no algorithm can decide whether that algorithm for a given input will either halt or go into an infinite loop.

Unfortunately for Turing, the proof that Hilbert's *Entscheidungsproblem* was unsolvable had been proved by the American logician Alonzo Church. Using his own analysis of a computable or effective procedure, a cumbersome formalism known as the lambda-calculus, Church proved that there could be no procedure for deciding whether an arbitrary statement in formal logic was valid or not. This is known as Church's Theorem.

Nevertheless, Turing's paper was published in the *Proceedings of the London Mathematical Society*. His professor, Max Newman, wrote a letter arguing that Turing's paper should be published not for the result

---

Church had been invited to give a talk at Stanford's Center for the Study of Language and Information (CSLI). Giving a tour of the facilities, John Etchemendy showed Professor Church a Dandelion computer whose operating system was based on functional programming which traces its roots to Church's Lambda Calculus (in contrast to the state-based programming languages based on Turing Machines). Before his talk, Church apologized for not showing more interest in the tour, explaining that he knew very little about computers, but that he once had a student who did–Alan Turing.[3]

but for the beauty of Turing's proof. Today Turing's paper is regarded as setting the foundation for modern computer science. Newman also wrote a letter for Turing to be a fellow at the Institute for Advanced Studies in Princeton. While at Princeton, Turing wrote a Ph.D. thesis "Systems of Logic Based on Ordinals" becoming Church's fourth Ph.D. student.

In the module *Critical Thinking as Computational Thinking*, we will learn more about Turing's remarkable work and how he built a computer to crack the Nazi Enigma Code during World War II, but it wasn't until 2009 that the United Kingdom publicly acknowledged that Turing and the codebreakers may have shortened the war by an estimated two years saving 14 millions lives. In September 2009 British Prime Minister Gordon Brown issued an apology, which in part, read:

> Thousands of people have come together to demand justice for Alan Turing and recognition of the appalling way he was treated.... This recognition of Alan's status as one of Britain's most famous victims of homophobia is another step towards equality, and long overdue.
>
> But even more than that, Alan deserves recognition for his contribution to humankind. For those of us born after 1945, into a Europe which is united, democratic, and at peace, it is hard to imagine that our continent was once the theatre of mankind's darkest hour. It is difficult to believe that in living memory people could become so consumed by hate—by anti-Semitism, by homophobia, by xenophobia, and other murderous prejudices—that gas chambers and crematoria became a piece of the European landscape as surely as the galleries and universities and concert halls which had marked our European civilisation for hundreds of years.
>
> It is thanks to men and women who were totally committed to fighting fascism, people, like Alan Turing, that the horrors of the Holocaust and of total war are parts of Europe's history and not Europe's present, so on behalf of the British government, and all those who live freely thanks to Alan's work. I am very proud to say: we're sorry. You deserved so much better.

After the release of the *The Imitation Game* [2014] staring Benedict Cumberbatch and Keira Knightly this part of Turing's life became more widely known. Turing was just 41 when he died from ingesting an apple laced with cyanide on June 7, 1954. For decades, his status was unknown due to the secrecy around his code-breaking work during World War II and the social taboos about his sexuality. In 2019 the United Kingdom announced that its new 50-pound note will bear a photo of Turing and will include a quote: "This is only a foretaste of what is to come and only the shadow of what is going to be."

Future Exciting Episodes

Time Magazine's Millennium issue of the 100 most important intellectuals and scientists of the 20th century lists Kurt Gödel, Ludwig Wittgenstein, and Alan Turing. Why does the esoteric, and relatively unknown, discipline of mathematical logic have such a disproportionately high representation? Gödel was the greatest logician of the 20th century. Wittgenstein was the greatest philosophical genius of the 20th century, and Turing was the greatest computer scientist of the 20th century. What could they learn from each other if they had to combine their intellectual ideas and life experiences to solve three puzzles to escape from a puzzle room? What would the 21st century reveal about the prescience of the work of these three geniuses?

Summary of Concepts

COMPUTATIONAL THINKING is a set of problem-solving skills that arose from thinking about the nature of computation. This skill set includes technical skills as well as designing human interfaces, and creating environments conducive to collaboration. Computational thinking will be as important in the 21st century as reading, writing and arithmetic were in earlier centuries.

ALGORITHMIC THINKING is the ability to set forth a solution to a problem by giving a set of explicit instructions, requiring no ingenuity, which is guaranteed to arrive at or calculate a solution. (To say the an algorithm requires no ingenuity is to say that it can be follow "blindly" or "mechanically" by a machine or computer in a way that doesn't require thinking!)

REVERSE ENGINEERING is the process of figuring out how an algorithm, given to you as a "black box," works or to discover its weaknesses or limitations.

DEBUGGING is the process of testing or verifying that your algorithm is doing what it is supposed to do in a systematic way.

BINARY CODING is the assignment of binary numbers or strings to the steps of an algorithmic process in order to provide a simple way of saying and seeing what is happening during that process.

DATA ARCHITECTURE refers to designing the data base (e.g., a spreadsheet) in such a way that it is easy to input, systematically store, and recall the information or data.

Some Tips on Performing Magic

The idea of using magic tricks to explain mathematics began when a student showed me some of her favorite mathematical card tricks. After that I began to be intrigued about what magic could teach us about ourselves. Misdirection, for example, isn't about moving your hands faster than the eye can see. Sometimes it's about how the human perceptual system "fills in the background" while using its information processing "pixels" to focus on something. An average cell phone has far more pixels for taking photos than the human visual system has rods and cones for perceiving. Mentalism is also a part of magic that is quite fascinating. It is amazing to me how susceptible the mind is to suggestion, and how many cures you can read from a person processing information. Here are some tips to keep in mind while performing the magic tricks.

"Any sufficiently advanced technology is indistinguishable from magic."
—Arthur C. Clarke (1917-2008)

- Humans aren't as good at following algorithms as computers. Aim at making your instructions as clear and memorable as possible. Moreover, watching someone perform an algorithm can get monotonous rather quickly, so you should have some patter to fill in.

- Practice before you perform the trick. Nothing is more annoying that a magic trick that's not executed properly.

- Make the trick your own. You can get creative about creating you own patter or story or making your reveal more entertaining. Consider the Middle Thirds card trick.

    # At the end you can deal the cards alternatively face down and face up, and the chosen card will be in the middle.

    # Or you can mix the cards facedown into random piles while slipping the middle or 14th card into your lap. Then add a little misdirection. While you are instructing the volunteer to guess what pile her card is in, it is surprisingly easy to place the chosen card in an unexpected location. Magic tricks are ways of "hacking" the human perception system, which can be an entertaining research tool.

    # Another possibility is to ask the volunteer to choose a denomination of bill that has a portrait of a President. (Note that the Alexander Hamilton ($10) and Benjamin Franklin ($100) were not Presidents.) That leaves four choices. Since you know that the chosen card will be at the 14th card, it is easy to rig the deck. If George Washington is chosen, you can deal two cards off the top, saying "one dollar." If Andrew Jackson or Ulysses S. Grant are chosen, you can simply slip a card from the bottom and put it on top before you begin spelling out the name. Honest Abe requires no finagling.

- It is useful to have alternative ways of performing a trick. This keeps the trick from getting old. A magician once fooled Richard Feynman. Every time Feynman would try to guess how the trick worked, the magician would perform the trick in a way to rule out Feynman's conjecture. Feynman didn't suspect that the magician was using different methods for achieving the same effect. Systematically, ruling his conjectures out one by one, didn't ensure that Feynman hadn't already guessed how the trick worked.

- Magicians need not be mathematicians: you don't have to understand an algorithm for it to work! However, magic can entice you to want to understand the mathematics, and so can be an effective pedagogical enticement to learn the mathematics to reverse engineer the trick. The more ways you have of executing an algorithm and more ways you have to comprehend how the algorithm achieves it magic.

- Magic can be a way of enticing someone to learn. Magic can begin in delight can end in mathematical wisdom.

    # For example, the Middle Thirds Card trick can be used to talk about Cantor's Discontinuum, a fractal formed by removing the middle thirds of a line segment:

    # Cantor's discontinuum has intriguing mathematical properties: it is everywhere discontinuous, it has the cardinality of the continuum and is uncountably infinite, it can be generated by a stochastic or coin-flipping algorithm, it is related to the fractal known as the Sierpinski triangle, and it has a *fractal dimension* $d = log(2)/log(3) \approx 0.631$.

- Have fun sharing the magic: the more it's shared and more it grows. But don't give away the secret until you've thoroughly engaged someone's curiosity. It is more fun, and educational, to let the person figure it out for themselves.

Exercises

Group I

1. *The Dating Game Algorithm*

Today our lives are increasing run by algorithms. Even dating has succumbed to the allure of the binary to calculate a potential partner with the most harmonious fit. The following algorithm, based on a card trick created by magician and computer programmer Alex Elmsley will introduce us to the usefulness of coding information using binary numbers.[6]

As mentioned above automatic card tricks are algorithms, so by playing around with them we can actually get a feel for other aspects of computational thinking–e.g., how to code data so it's easy to track what an algorithm is doing, how to debug algorithms, and how to test the limits or break algorithms.

Our version has two individuals enacting the algorithm simultaneously to keep it more interesting. In magic circles, Emsley is known for the "Ghost Count" or "Emsley Count" which depends on manual dexterity. Fortunately, this automated card trick depends only on following an algorithm.

The Set-Up:

1. The Magician shuffles the deck, counts out two face down decks of 16, places two Jokers face-up on the decks, and fans the remaining cards face down between the two decks. Then the Magician asks for two volunteers—whom we shall call LEFT and RIGHT, each of whom has one of the piles with Jokers placed in front of them.

2. The Magician asks LEFT and RIGHT each to choose a card from the middle array of fanned out cards. This is their secret of identity card for the Dating Game. "Remember it or you'll never get a date and the trick won't work. Do you have it memorized?" The Magician instructs LEFT and RIGHT to place their secret identity cards at the bottom of the decks and to discard the Jokers.

3. The Magician asks RIGHT to divide the remaining middle array into two piles anywhere she wishes (but making sure that the division of the 20 remaining cards leaves no pile has less than 5 cards). Then the Magician asks LEFT to choose one of the two piles and to place his deck on top of it. RIGHT places her deck on the remaining middle cards. The Magician asks RIGHT and LEFT to exchange decks.

The Trick:

4. Ask RIGHT and LEFT to stand back-to-back so they can't see that the other is doing. "Now each of you is going through four rounds of speed dating. "Now dating can have its ups and downs, right? So, alternately deal your cards *face-down* in one pile and *face-up* in the other. Continue in this manner until all the cards have been dealt out.

5. "Dating can have a lot of downers, so eliminate the face-down cards. Take the face-up pile, turn it over.

6. Repeat steps 1 and 2, until you're left with one card.
7. Then place the last card on your chest face down and turn around and face each other.

The Reveal:

8. Ask RIGHT and LEFT to reveal their secret identity cards.
9. RIGHT and LEFT then over the cards on their chests.
10. The binary algorithm has selected them for each other.

The set-up places the secret identity cards to be at the 16th position from the top of each deck. If you start numbering the decks with binary numbers starting with 0, the 16th position has the code for 15 = 8 + 4 + 2 + 1 is 1111 in binary.

(A) Explain why the trick works using binary coding.

(B) The Dating Game Trick does not work if one of the piles from the Middle has less than 5 cards. Explain why using the binary coding.

(C) Reverse engineer the trick to that you, by adding an additional number of cards to one of the decks, have to algorithm choose the Joker.

2. *The 16-16-11 Card Trick*

This trick was taught to me by one of my students, Sarah Eller, who learned it from her father. The goal of this exercise is to learn how to explain why the trick works using binary coding.

A Magician deals out a deck of 52 card into three piles of 16, 16, and 11 all face down. Three volunteers are chosen to select a secret card from each of the piles and then, after memorizing the card, to place it back top of the pile. The Magician "stacks"* the deck. With the deck face-down, she begins dealing the cards alternatively face-up and face-down into two piles. The Magician asks the volunteers to take note of when their chosen cards appear. After removing the face-up pile, the Magician repeats this procedure with the remaining face-down deck until there are three cards left.**

*Instructions for "stacking the deck:" When the cards are original dealt into three pile 9 cards remain to the side. When "stacking the deck," the Magician places those 9 cards on top of one of the piles of 16, then places those 25 cards on top of the next pile of 16, and then places all of those cards on top of the remaining 11.

Finally, the Magician asks the volunteers: Whose card showed up on the first deal? On the second deal? On the third deal? On the fourth deal? Then the last three cards are turned over to reveal the chosen cards.

**On the third deal, the Magician secretly places the top card on the bottom of the deck before dealing out the cards.

Explain why the trick works using binary numbers.

Group II

3. *The Middle Thirds Card Trick*

This trick involves 27 cards from a standard deck.

The Set-Up:

1. Arrange 27 cards into 3 rows of 9 cards each. Deal them vertically in the order as shown in the diagram. (When you deal the cards for each row, slightly overlap the cards so it will be easier to gather up the cards horizontally in order.)

000	010	020	100	110	120	200	210	220
001	011	02	101	111	121	201	211	221
002	012	022	102	112	122	202	212	222

2. Ask a spectator to mentally choose one of the cards and to point to the row in which is appears.
3. Gather up the cards in each row into three separate piles and reassemble the deck of 27 cards making sure to place the pile with the chosen card in the middle of the deck and to keep those cards in order.
4. Repeat steps 1 – 3, three more times.

The Reveal:

5. Deal out the cards alternately face down and face up.
6. The chosen card will be face down in the exact middle of deck–i.e., it will be the 14th card from the top (starting the count with 1).

Use the ternary (base 3) coding to explain why the trick works.

Group III

4. *The I Ching Colorized*

When the German philosopher Gottfried Leibniz (1646-1716) first saw the 64 hexagrams used in the ancient Chinese art of the I Ching, he thought that the Chinese had discovered binary numbers.

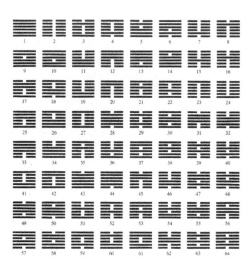

This is a version of the I Ching that uses a deck of playing cards instead of yarrow sticks. The "Colorized I Ching" combines fortune telling tricks from astrology and a little magic from mathematics. The purpose of this exercise is to learn a little about the fallacies of pseudo-science and about the mathematics of binary numbers.

1. Deal the deck of 52 cards into piles of 30, 12, and 10.
2. Choose a card from the deck of 10 and add it to the deck of 30. If the card is 7, for example, then deal the deck of 30 cards into two equal piles of 7 cards each. This number can also be linked to the number of the volunteer's birthday. If the number is 1, 2, or 3, you can add 10 to make it 11, 12, or 13.
3. Next ask the volunteer to choose secretly one of the 12 cards, arranged like a clock, and memorize it. This choice can be based on the person's astrological sign, birth month, etc.
4. Ask the volunteer to place her secretly chosen card on top of one of the two dealt piles.
5. Drop the remaining cards from the deck of 30 on top of the secret card.

6. Now take that newly assembled pile and deal "down-and-under" until there is only one card left.
7. This is the secretly chosen card.

Based on the binary numbers, this trick works with any deck of cards with 2^n cards. Explain why.

5. *Digital Computing*

Finger multiplication can be generalized to the half-decade 11 – 15 larger than 10. The fingers are labeled 11 (beginning with the pinky) through 15 (ending with the thumb).[7]
To multiply 13×14, touch the two corresponding fingers and use the following algorithm:

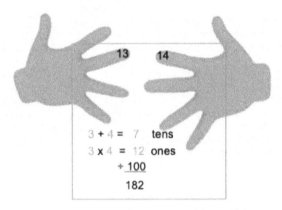

$$13 \times 14 = 70 + 12 + 100 = 182$$

This time *add* the fingers touching and below to obtain the *tens*.
But *multiply* those same fingers to compute the *ones*.
Then *add* a *constant* of +100.

Explain why this algorithm works both algebraically and geometrically.

6
Cultivating Creativity

Life is not easy for any of us, but what of that? We must have perseverance, and above that, confidence in ourselves. We must believe that we are gifted for something, and that this thing, no matter what the cost, must be attained.[1]
— MARIE CURIE (1867-1934)

A recurrent theme of this module has been that creative problem solving is not an unlearnable skill that is accessible to only a select few geniuses. Problem solving and creativity can be enhanced with a toolbox of heuristics, but they are also habits of mind that can enhance our lives. In this concluding chapter, we'll discuss some limitations of the models we have been using to explore the elements of problem solving. Then we'll suggest some activities to add to your repertoire of problem solving skills and suggest ways to approach life in ways that cultivate creativity.

MARIE CURIE

Some Problems with Puzzles

Our framework for understanding problem solving was popularized and developed by Alan Newell and Herbert Simon in *Human Problem Solving*.[2] Here are the key conclusions of their research program:

- Problem solving can be viewed as a selection of a *solution path* from a space or tree of alternatives.

- Due to the limits of short-term memory, we tend to narrow our focus by working our way through a problem space in *serial* fashion, trying one path at a time rather than *simultaneously* searching the problem space in *parallel* fashion,

- *Models* help us to overcome the limits of short-term memory and make it easier break down the problem into pieces so that we can conduct *simultaneous searches* in disconnected parts of the problem space to find a solution path.

- A major characteristic of human problem solving is that we use *selective*, not exhaustive, searches. We need to develop a sense when this selectiveness is the result of expert experience that facilitates efficiency and when that selectiveness prematurely prevents us from exploring other, perhaps more fruitful, lines in inquiry.
- Problem solving can be improved by learning *heuristics*, rules of thumb for pruning the tree of alternatives thus making it easier to navigate through the problem space in more efficient ways. Heuristics, however, are no substitute for creative and critical experience with the subject matter at hand.

If this is to be a book that *explores* critical thinking and itself *engages* in it, it is incumbent upon us to think critically about the model of problem solving that Newell and Simon have handed us. The information-processing model of problem solving proposed by Newell and Simon is now commonplace in artificial intelligence (AI).

One clear advantage of this model is that it gives researchers a way of testing hypotheses about how human beings solve problems. We can simulate human problem solving using AI thus generating testable predictions. Are there important and relevant *dissimilarities* between how computers and humans solve problems?

Ironically, one of the results that has emerged from the intensive study of human problem solving using computer modeling is the *computational paradox*. Computer modeling of problem solving has revealed, often quite precisely, how human problem solving differs from how problems are solved using efficient computer algorithms. Moreover, attempting to show how creativity can take place by orchestrating such ordinary activities as *noticing, choosing, valuing*, etc., hasn't diminished the mystery of creating. It has only relocated those mysteries in the ordinary activities we so often take for granted–for example, how creative persons *notice* critical, often overlooked, features of a problem. Paradoxically, complex computational problems like playing chess have proven to be relatively easy to solve while modeling everyday mental processes and schemas have often proven to be quite elusive.

One of the consequences of using simple computational models of problem solving has been to focus research on relatively artificial puzzles–classic recreational logic puzzles such as the Towers of Hanoi, crossing the river with a wolf, sheep, and cabbage, or using a balance to find the one counterfeit coin (e.g., a classic puzzle is to find the counterfeit coin which may be lighter or heavier among a dozen otherwise identical coins in the fewest weighings). Such puzzles are useful for research precisely because they don't require a great deal of additional real world information. They provide a *controlled* way of studying some aspect of problem solving without having to supply common sense background information.

However, a theory based on how people solve puzzles and play parlor games need not be a good theory of how we solve problems "in real life." Using the research on puzzle solving to explain human problem solving has been compared to the proverbial story of the drunk looking for his keys under the streetlight. When a policeman asks the drunk why he was looking only under the streetlight when he could have dropped his keys anywhere between the lamppost and the bar, the drunk responds in exasperation, *"cause it's the only place I can bloody well see!"* One might say something similar about the research into human problem solving based on solving puzzles: perhaps it is where the light is, but it may not be where the keys to human problem solving in "real life" are to be found.

Another limitation of the use of puzzles as research tools is that they are *knowledge-poor*. Unlike most problems we encounter, the puzzles usually require no specialized knowledge about the world. Acknowledging this problem, many AI researchers devoted their time to developing 'expert systems' in specific domains. These expert systems try to solve a problem in a limited domain, not by using generalized heuristics, but by incorporating the context-specific heuristics experts have acquired through experience.

As a sophisticated joke, Joseph Weizenbaum of MIT wrote the ELIZA program (named after the character "Eliza Doolittle" in George Bernard Shaw's *Pygmalion*) to mimic the responses of a Rogerian psychotherapist talking to a patient. Carl Ransom Rogers (1902-1987) was known as the founder of "client-centered therapy" in which the therapist "reflects" back the client's words. To "sidestep the problem of giving the program a data base of real-world knowledge," Weizenbaum programmed the AI to simply pick out key words or punctuation marks in the patient's typed-in responses to trigger replies that appeared to be those of a Rogerian therapist.[3]

```
woman: Men are all alike.
eliza: In what way?
woman: They're always bugging us about something
       specific or other.
eliza: Can you think of a specific example?
woman: Well, my boyfriend made me come here.
eliza: Your boyfriend made you come here?
woman: He says I'm depressed much of the time.
eliza: I am sorry to hear you are depressed.
woman: It's true. I'm unhappy.
eliza: Do you think coming here will help you not to
       be unhappy?
woman: I need some help, that much seems certain.
eliza: What would it mean to you if you got some help?
woman: Perhaps I could learn to get along with my
       mother.
eliza: Tell me more about your family.
```

How did ELIZA work? The human communicated with ELIZA through teletype. When ELIZA received text ending with a '?', it responded: "Why do you ask?" When ELIZA received a text containing "might" it responded: "Why aren't you sure?" ELIZA replied to sentences with pronouns such as "I think you don't like me" with the question: "What makes you think I don't like you?"

Weizenbaum eventually became alarmed about how easy it was to fool humans:

> The whole issue of the credibility (to humans) of machine output demands investigation. Important decisions increasingly tend to be made in response to computer output. The ultimately responsible human interpreter of "What the machine says" is, not unlike the correspondent with ELIZA, constantly faced with the need to make credibility judgments. ELIZA shows, if nothing else, how easy it is to create and maintain the illusion of understanding, hence perhaps of judgment deserving of credibility. A certain danger lurks here.[4]

Weizenbaum lamented: "I had not realized ... that extremely short exposures to a relatively simple computer program could induce powerful delusional thinking in quite normal people."

As we have tried to emphasize earlier in the *haiku* example, you can't automatically become a master at any given field by just knowing a few heuristics, even domain specific heuristics. There's no substitute for training and experience. Research shows that experts tend to have tailor-made heuristics—quick little associations, based on lots of experience, linking one or two scraps of information with the heart of the problem—pointers, crosslinks in the knowledge network.

But these specifically tailored heuristics are often domain specific and can be readily applied only if one has had plenty of experience.

Extending Memory Beyond Miller's Magic Number

Consider some of the research on improving memory. Remembering can itself be viewed as a special sort of problem solving. Psychologist George Miller in a famous paper, "The Magical Number Seven, Plus or Minus Two: Some Limits on Our Capacity for Processing Information," discovered that most people can conveniently only remember about 7 ± 2 bits of unrelated information.[5]

Researchers have found that people can increase the capacity of their short term memory by *chunking* things together in smaller groups. For example, many people find it difficult to memorize sequences of random letters:

KRARUOYDLIUBOTSNIARTILITNUTIAWTNOD

Although we may all know a person who can memorize the list at a glance, for most of us it is helpful to chunk the letters into memorable *clusters*. (Viewing the sequence *backwards*, break the sequence into a series of easy-to-remember clusters).

Another way to increase our memory capacity is to code items using a peg-word system. Suppose we wanted to memorize the following sequence of nine words:

BRIM, HOT, WASP, WOES, HEAR, TIED, TANK, SHIP, FORM

To use a peg-word system, create a mental picture with a sequence of prearranged items—or pegs. The pegs might be based on a simple rhyme such as the following on the right.

After memorizing this list of 'pegs,' you can remember other items by associating or 'hanging' the items to be remembered on each of the 'pegs'. To do this, you use your imagination to combine the peg and the item to be remembered into one vivid image. The more unusual and absurd the image or association the easier it is to remember.

Here, for example, is one way of using the peg-word system.

ONE-BUN	A BUN filled to the BRIM with butterflies.
TWO-SHOE	A SHOE covered with HOT sauce.
THREE-TREE	A TREE pruned to look like a WASP.
FOUR-DOOR	A DOOR for widows with WOES.
FIVE-HIVE	A buzzing BEE HIVE you can HEAR in your ear.
SIX-STICKS	A bundle of STICKS TIED with a bow.
SEVEN-HEAVEN	HEAVEN emptying a rainwater TANK.
EIGHT-GATE	A GATE adorned with a SHIP in a bottle.
NINE-SIGN	The SIGN with a nonsensical FORM.

ONE		BUN
TWO		SHOE
THREE		TREE
FOUR		DOOR
FIVE		BEEHIVE
SIX		STICKS
SEVEN		HEAVEN
EIGHT		GATE
NINE		SIGN
TEN		HEN

With this peg-word system coding the random words, it should be easier to recall the items. For example, try to list the odd-numbered items in descending order followed by the even-numbered items in increasing order. I've enjoyed teaching many students and friends how to make this system their own to pass tests requiring rote memorization–ranging from Abraham Maslow's stages of actualization to the functional systems of the cerebral cortex.

The *memory palace* is another famous technique for enhancing memory, a technique taught by the Jesuit Matteo Ricci (1552-1610) in the imperial courts of China in his missionary efforts to foster dialogue between the Pope of Rome and the Emperor of China. You can visualize a walk through your home—or memory palace—placing objects to remind you of your marks. As you approach the front door of your place, there is a poster of an ear on it: "*Friends, Romans, countrymen, lend me your ears.*" As you enter the living room, you see a grave: "*I have come to bury Caesar, not to praise him.*" The room of filled with various conspirators—whom you call with irony "honourable men."

As you enter the kitchen, there is a pot cooking on the stove. The post is labelled "ambition." When you take off the lid of the pot and peer inside, you see a crown of the poor: "*When the poor have cried, Caesar hath wept: ambition should be made of sterner stuff.*" On the kitchen counter are three crowns—signifying that many have seen Caesar refused the crown, or coronet, three times. Retreating to the bedroom, you see a coffin: "*My heart is in the coffin there with Caesar/And I must pause till it come back to me.*" Next to the coffin is Caesar's will, which do you not dare to read since that would "*wrong the honorable men whose daggers have stabbed Caesar.*" Entering the bathroom, you see Caesar's body in the bathtub. You point out his wounds one at a time made by the "honorable men," especially the stab wounds made by Brutus, ("*Judge, O you gods, how dearly Caesar loved him!*").

Going out the back door into the yard, contrast Brutus "an orator" with the words of the "plain blunt man." Only one as eloquent as Brutus could give voice to each of Caesar's wounds: "*...that should move/ The stones of Rome to rise and mutiny.*" Addressing the crowd in the backyard read Caesar's will: "*To every Roman citizen he gives,/ To every several man seventy-five drachmas*" as well as land. As the moon rises: "*Here was a Caesar, when comes such another?*"

Experts in their given field, however, tend to use an even more effective memory strategy: they learn more, and more deeply. A children's book on Native Americans, for instance, listed the following information:

> In the Southwest, the Navajo Indians lived in adobe houses. In the Northwest, the Indians lived in wood houses with slanted roofs. And in the Midwest, Indians lived in tepees and hunted buffalo.[6]

If you were a child trying to remember this, you might try some peg-word system. But an easier way is to relate the bits of information to a larger set of information. Suppose you know that in the Southwest, there's a lot of hot sun and mud. Imagine the Navajos baking adobe bricks. What do you know about the Northwest? Perhaps you know that it rains a lot and there are a lot of trees. What would be the natural material with which you could build a shelter? You might want to build wood houses with slanted roofs to protect you from the rain. The nomadic Native Americans of the Plains needed shelters that were mobile in order to follow the buffalo herds. What kind of shelter is easily put up and taken down? Tepees. When the isolated pieces of information can be unified into one coherent story, the information becomes easier to recall, reconstruct, and remember.

Research into problem solving has also shown that experts in a given domain tend to construct *hierarchical* knowledge networks. Experts have a high-altitude overview and as a result they have enormously efficient pictures of just what they're trying to do when solving a problem. They see things in perspective, so they know just what additional information they need. In contrast, beginning students tended to solve physics problems, to collect or compute a great deal of unnecessary information and to plug that information into formulas blindly. They didn't know what they really *needed* to know.

Expert knowledge depends not only on knowing specific content, but also on recognizing formal or structural similarities between seemingly unrelated problems. Besides having more experience (and consequently better ways of remembering and learning), experts tend to have and maintain higher critical standards–they notice more, remember more, exercise better critical judgment, and maintain higher standards. Critical thinking involves style, values, beliefs, and what philosopher Israel Scheffler (1923-2014) has called "cognitive emotions." Scheffler realized that cognition and affect need not be distinct aspects of creative experience:

> A love of truth and a contempt of lying, a concern for accuracy in observation and inference, and a corresponding repugnance of error in logic or fact. It demands revulsion at distortion, disgust at evasion, admiration of theoretical achievement, respect for the considered arguments of others.[7]

Cognitive emotions can provide knowledge or they can point to knowledge. Emotions are a way of knowing.

Current research on problem solving tends to ignore the social dimensions of creativity. Today the world is facing more complex problems and greater cultural diversity than ever before. A global renaissance of creative problem solving may be our best hope for a better world.

The conditions for social creativity have been explored by Dean Simonton at the University of California. Scrutinizing 127 twenty-year periods in European history from 700 BC to AD 1839, Simonton found that political fragmentation was the single best predictor of creativity. Creative development, in other words, depends on exposure to political change and cultural diversity. Simonton found that throughout history, intense rivalries among small states have often sparked creativity.

Unreasonable Effectiveness of Logic

Modifying a memorable phrase of Nobel physicist Eugene Wigner (1902-1995), we may ask what accounts for the "unreasonable effec-

The miracle of the appropriateness of the language of mathematics for the formulation of the laws of physics is a wonderful gift which we neither understand nor deserve.
—Eugene Wigner

tiveness" of logic? I want to explore this question by presenting two games.

A game called "Hot" can be played with the nine words you have memorized. Here again is the list in alphabetical order:

Brim, Form, Hear, Hot, Ship, Tank, Tied, Wasp, Woes

These words are written on nine cards. Two players take turns choosing cards. The first player to recognize that she has three cards with words containing a common letter, e.g., Hot, Tied, Tank wins the game. Please take time to play Hot a few times to get a feel for how to play.

A game called "Ace" is played with the cards Ace – 9. Again two players take turns choosing the cards. The first player to obtain three cards whose sum is 15, e.g. $4 + 5 + 6 = 15$ wins the game. Please take some time to play Ace a few times.

After a while, the two games may start to feel similar. With best plays by both sides, the games end in a draw. To see the strategic value of knowing a mapping, here's a simple puzzle. Suppose player 1 chooses "Form" and player 2 responds with "Woes", can you find a winning move for player 1?

Playing Hot and Ace is worthy of closer examination. Which version of the game did you find easier to play? For those who don't like adding and subtracting, perhaps Hot is preferable. For many, however, Hot is harder to play because it is tedious to keep checking for letters in common. Perhaps you found Ace to be more manageable because it's easier for you to compute when the sum of three numbers is 15, or to subtract from 15 in order to figure out how to block your opponent from getting a winning triple.

Actually, the two games are more than just similar: they are, mathematically speaking, *equivalent*. In the mathematician's jargon the two games are *isomorphic*. There are several (symmetrical) one-to-mappings that preserve all the winning (and losing) relationships. One such mapping is the one you have coded by means of the peg-word system.

Brim	Form	Hear	Hot	Ship	Tank	Tied	Wasp	Woes
Ace♥	9♣	5♠	2♦	8♦	7♥	6♣	3♣	4♦

Notice that whenever three words in Hot have a common letter, their corresponding cards in Ace add up to 15, and whenever three cards in Ace sum to 15, their corresponding words in Hot have a common letter.

Actually, you've probably discovered that both Ace and Hot are isomorphic to a game with which you are quite familiar. This can be seen by mapping the pairings of numbers with words onto the "magic square":

2 BRIM	9 HOT	4 WASP
7 WOES	5 HEAR	3 TIED
6 TANK	1 SHIP	8 FORM

The magic square (also known as the Lo-Shu) gives you all the ways of finding three of the numbers from 1 - 9 that add to 15 in terms of the rows, columns, and diagonals. Winning HOT or ACE is equivalent to getting three in a row in TIC-TAC-TOE!

Knowing this isomorphism gives the first player a considerable advantage over a player who is initially unfamiliar with HOT or ACE. For example, if the first player chooses any of the corner squares, she is guaranteed to win with best play against a player who fails to choose the center cell. Correct play on both sides, of course, inevitably results in a draw.

Another way to play is *reverse* tic-tac-toe where both sides try *not* to get three in a row. This version leads to a draw with best play, but it's much harder for the first player to obtain the draw.

This miniature puzzle provides insight to why logic is so powerful–so "unreasonably effective" in solving all kinds of problems. Natural language can obscure logical relationships. Consider, for example, the following valid instance of *modus ponens*:

> If Socrates is in Athens, then he is not in Sparta.
> Socrates is in Athens.
> Therefore, it is not the case that Socrates is in Sparta.

Yet when we replace "Socrates" with "Somebody" we obtain:

> If *somebody* is in Athens, then he is not in Sparta.
> *Somebody* is in Athens.
> Therefore, it is not the case that *somebody* is in Sparta.

This is obviously a logical fallacy. (If somebody being in Athens could have vacated the inhabitants of Sparta, Socrates's heroism in the Peloponnesian War (431-404 BC) would not have been required!) Apparently, a proper name like 'Socrates' is logically different from a quantified expression like 'Somebody'. But why? Fascinating linguistic puzzles like these are discussed in the module, *Critical Thinking as Logic and Deduction*

When natural language harbors sophisms like this, our thinking can be clarified by translating natural language into a symbolic logic.

This discipline forces us to correctly represent the logical form of everyday expressions. When we transitioned from Hot to Ace we replaced the distraction of words with the utility of numbers. Logical deductions can, in principle, be reduced to number crunching. However, this level of detail is often too tedious for us to see the strategic clarity offered by Tic-Tac-Toe. The level of abstraction of symbolic logic seems just right: it has enough abstraction to extricate ourselves from the jungle of natural language, but not so much as to lose us in a jumble of minute computations.

Recall the isomorphism between the problem of Euler's Lake and Kant's Ghost revealed that they were essentially the same problem. In Chapter 3, *Bridges to Problem Solving*, we considered two seemingly very different puzzles. One involved traversing the Königsberg bridges in one continuous walk, the other involved breaking through the walls of an Egyptian maze:

 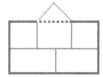

Euler's Lake with 16 Bridges. Starting from one of the five islands, can you plan a continuous walk that crosses each bridge only once? With the dotted bridge, is the puzzle solvable?

Kant's Castle. Starting from any room or from outside the castle, Kant's ghost must pass through each wall only once in one continuous path. By replacing the dotted wall with the two blue walls, is the puzzle solvable?

Another way of exploiting formal or structural relationships is to embed a problem into a richer context in which there is a robust set of conceptual tools ready to be used.

A ladybug crawls across two identical adjacent square tiles. One is perpendicular to the other. There are two paths, blue and dotted purple. What angle does the ladybug's path make in space?

Descartes embedded geometry into an algebraic language by describing the figures on what is now known as the *Cartesian coordinate plane*. Using the Pythagorean theorem, for example, a circle can now be described algebraically, as the set of all points (x, y) a given distance r from the origin (a, b).

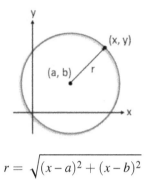

$$r = \sqrt{(x-a)^2 + (x-b)^2}$$

Exploiting this richer setting, 19th century scientists were able to solve three famous outstanding geometry problems posed by the ancient Greeks: the duplication of the cube, the squaring of the circle, and the trisection of an arbitrary angle. The problem was to make an exact construction with the aid of a compass and straightedge. It was shown that it is logically impossible, given the conditions of the problem, to construct a cube with exactly twice the volume as the original, to construct a square with exactly the same area as a given circle, and to trisect an arbitrary angle.

Wiles Waiting to Prove Fermat's Theorem at Last

Andrew Wile's proof of Fermat's Last Theorem [1995] is one of the most famous examples of using the dynamic interplay between algebra and geometry to solve a problem.

The story begins around 1637 when Pierre de Fermat (1607-1665) scribbled in the margins of his copy of Diophantus's *Arithmetica* that there are no solutions for $a^n + b^n = c^n$, for n greater than 2. Fermat claimed to have a proof that was "too large to fit in the margins." Generations of mathematicians wrestled with Fermat's problem for almost 360 years before Fermat's *conjecture* was turned into a theorem. This transformation was accomplished by the British mathematician Andrew Wiles, who devoted seven years of his life to this grand obsession.

Wiles's proof is by contradiction. If Fermat's Last Theorem is false, you can construct semi-stable elliptic curves that are *never* modular.

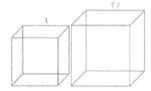

DOUBLING THE CUBE

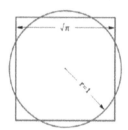

SQUARING THE CIRCLE

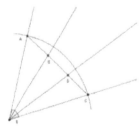

TRISECTING THE ANGLE

Theorem: There exists no $n > 3$ such that $\sqrt[n]{2}$ is rational.
Proof: Assume for *reductio ad absurdum* that $\sqrt[n]{2} = p/q$, for $n > 3$. Then $2 = p^n/q^n$ and so $2q^n = q^n + q^n = p^n$, for $n > 3$, which contradicts the Fermat-Wiles Theorem.
$3987^{12} + 4365^{12} = 4472^{12}$
—*The Simpsons*, Season 10, Ep. 2. "The Wizard of Evergreen Terrace"[8]

> *Perhaps I can best describe my experience of doing mathematics in terms of a journey through a dark unexplored mansion. You enter the first room of the mansion and it's completely dark. You stumble around bumping into the furniture, but gradually you learn where each piece of furniture is. Finally after six months or so, you find the light switch, you turn it on, and suddenly it's all illuminated. You can see exactly where you were. Then you move into the next room.... .* —Andrew Wiles

However, transposing statements about elliptic curves into their Galois representations, one can prove that these Galois representations are modular. This contradicts the original construction.

The details of Wiles's published proof required 129 pages. Wiles first announced his result in 1993 in a lecture at Cambridge entitled "Modular Forms, Elliptic Curves and Galois Representations" but his proof was found to contain a gap, which was finally repaired by working with one of his former graduate students and published two years later. Wiles was awarded the Abel Prize in 2016 for his "stunning proof."

Historically, times of crisis have been occasions for rethinking society creatively. Perhaps you've heard it said that the Chinese word for "crisis" (wēijī) is composed of two characters—wēi which means "dangerous" and jī which means "opportunity." This meme became popular around 1960 when John F. Kennedy introduced it in his presidential campaign speeches and was subsequently spread through New Age and business management literature. Actually, the second character is more accurately translated as "change point" leading to the memorable meme that, from a Chinese point of view, every crisis is an occasion or opportunity for change–or a "tipping point."

- Philosophy and western science were born in ancient Greece, the birthplace of philosophers as Socrates, Plato, and Aristotle. Ancient Greece was known for its numerous and competing city-states.
- During the Italian Renaissance, the bubbling political intrigue between rival cities such as Florence and Venice produced the geniuses Michelangelo, Raphael, Dante, Machiavelli, and Leonardo Da Vinci.
- When Germany was comprised of a mosaic of small principalities it offered to the world Mozart, Beethoven, Goethe, Hegel and Schiller. When Bismarck unified Germany in the late 19th century, the German Golden Age drew to a close. As Gladstone, the historian, put it: "[Bismarck] made Germany great and Germans small."
- The most recent Golden Age of creativity in America peaked after World War II. Simonton notes that "we began to be highly creative in all domains, becoming world leaders. Most of the world looked to us for leadership in all the various fields of science, of art, of the humanities in general." One reason for this creative spurt, Simonton notes, was America's ability to draw upon the diversity of its own people, whether refugees from Europe or African-Americans.

> *Historically, pandemics have forced humans to break with the past and imagine their world anew. This one is no different. It is a portal, a gateway between one world and the next. "We can choose to walk through it, dragging the carcasses of our prejudice and hatred, our avarice, our data banks and dead ideas, our dead rivers and smoky skies behind us. Or we can walk through lightly, with little luggage, ready to imagine another world. And ready to fight for it.* —Arundhati Roy "The Pandemic is a Portal"

Are the competitive start-ups of today similar to the rival city-states of the past? Western economies were more innovative than the old massive, centralized Eastern bloc economies because they allowed room for innovation. America's history is rich in creativity, in part because of its ethnic and racial diversity. What is happening in America today with regards to actions taken against diversity and immigration? What effect might this have on America's reputation for ingenuity or the future of innovation in America's businesses?

Beyond the Brain's Illusions

Critical thinking has been thought of as the avoidance of making mistakes–"elusion of cognitive illusions." Focusing on avoiding mistakes, rather than feeling free to make mistakes early and often, but learning from them, can lead to "analysis paralysis."

Another way of approaching critical thinking is to see what we can learn from our mistakes–especially from our brain's susceptibility to *systematic* cognitive mistakes. Perhaps these "mistakes" reveal ways that our cognitive abilities actually function quite efficiently in helping us navigate our way through life.

Why do visual illusions fascinate us? Philosophers, even going back to before Plato, have argued that illusions are evidence that our senses deceive us and so cannot be bases of certain knowledge. The persistence of an illusion, however, can lead us in another direction. Instead of eschewing error, we can pay attention to what it is teaching us. The quantification of error, for example, is the basis of statistics, which can be a valuable guide to life. The persistence of an optical illusion can be evidence of how our perceptual systems have evolved to help us navigate the world.

Consider Edward Adelson's Checker-Shadow Illusion. This illusion can be found online in many places (which is itself testimony to our fascination with illusion). Are squares A and B the same or different shades of gray?

ADELSON'S CHECKER-SHADOW ILLUSION

Why do we persistently perceive A is darker then B (*left*) even when the shades of gray are shown to be the same (*right*)? This persistence, Adelson argues, is evidence that our visual perceptual system does not function like a light meter, but rather as a *computational system* that evolved for navigating through the world. Moving through the world requires us to take into account such features as "*local contrast*" (lightened squares surrounded by darker squares and darkened squares surrounded by lighter squares), "*snap-to-grid*" (ignore gradual changes in light level when compensating for shadows), and "*paintiness*" (when a checkerboard pattern occurs, interpret differences of color as due to the surface colors of the squares). When all three computational corrections come into play the illusion is robust.

Optical illusions, like magic tricks, fascinate us because they "hack" our perceptual system. What the stubborn *persistence* of the illusion teaches us is not why our perceptual system *fails*, but why it *succeeds* so well in interpreting the color of objects in the world most of the time.

The "Spinning Dancer" is another fascinating illusion created in 2003 by Japanese web designer Nobuyuki Kayahara. The silhouette of a pirouetting dancer appears to be spinning clockwise to some and counterclockwise by others. However, when you attempt to reverse the spin (say by concentrating on the feet) then all of a sudden the dancer appears to reverse her spin. This illusion, too, is quite persistent.

—Nobuyuki Kayahara

Creating Lives

Hopefully this module on problem solving has put you in touch with your child-like joy in finding creative solutions to problems. There are no simple recipes for creating lifestyles and communities of creativity, but here is a real life example of "problem solving" and a testimony of what a caring community of students can accomplish during a class on Logical and Critical Thinking.

Compassion is also a creative way of knowing. One of the joys of teaching critical thinking is that students who catch the spirit of the course will often volunteer to come back and help teach others. The following excerpt is from an article "Euthanasia: A Personal Perspective" that I wrote for a newsletter for the Department of Preventive Medicine, *Contexts: A Forum for the Medical Humanities* (vol. 2, no. 2, Dec. 1993).[9]

Every man must decide whether he will walk in the light of creative altruism or in the darkness of destructive selfishness. This is the judgment. Life's most persistent and urgent question is 'What are you doing for others?'
—Martin Luther King, Jr.

> Euthanasia, I thought, would be a good issue to grab the attention of my philosophy students. How we answer the question "When is it right to die?" will shape the kind of society we will be living in for the coming decades. Unlike other moral issues (e.g., abortion or the death penalty), euthanasia is one of those issues which students are still open to thinking about. About 30 years ago the issue of euthanasia centered around the issue of needless suffering and pain. Today, when we know much better how to medicate for pain, the debate has shifted to "self-deliverance" for those despairing of living.
>
> I invited a colleague Dr. Peter Williams to lecture to my class. Dr. Williams required my unsuspecting class of 170 students to see a research video "Let Me Die." This video graphically records the agonizing treatment of the famous Texas burn victim, Donald C., who insisted on his right to die. Realizing that he could no longer live the life of a rodeo performer and ladies' man, Donald C. pleaded to be released from his painful treatments with an overdose of heroin.

After seeing this video before the class, many students were all too ready to support physician assisted suicide. Curiously, this readiness remained even when the students were later told that Donald C., after being granted permission to refuse treatment, suddenly had change of heart and decided to continue his treatments. Donald C. in fact went on to law school, married, and then gave lectures on the right to die.

In the Socratic tradition, Peter Williams provoked my students to offer their opinions and then, in his inimitable and inimical way, proceeded to sting them into the awareness of their own ignorance and inconsistencies. Pedagogically the lecture was a tremendous success, yet I was still disturbed. Why?

There are two kinds of intellectual obstacles. There are those rational and logical obstacles that can be overcome with argument, and then there are those issues that the modern mind simply does not see with its narrowed vision. We are all part of modernity. Even when we criticize it, we are influenced by it—like the family we grew up in, it remains part of us.

Modernity elevates the myth of the autonomous self. As Robert Bellah et al. pointed out in *The Habits of the Heart: Individualism and Commitment in American Life*:

> There are truths we do not see when we adopt the language of radical individualism. We find ourselves not independently of other people and institutions but through them. We never get to the bottom of our selves on our own. We discover who we are face to face and side by side with others in work, love, and learning.[10]

Why should my students think it so "painfully" obvious that the primary, if not the sole, issue was that of respecting the patient's autonomy? "If anyone autonomously chooses to end his or her life, then, of course, they have the right to end their life and they also have the right to have a physician assist them."

Why didn't they think to question whether this principle was really true, or whether the principle would, at least, have limited application in the case of a depressed person who had just undergone burns of this severity? Didn't Donald's sudden change of heart suggest that what he wanted was not to be allowed to die but to have some sense of control over his own life and perhaps to have a hope for a meaningful future? Why didn't they ask whether other burn victims were later grateful that their anguished requests to die were not respected? Why didn't they ask whether Donald C. could have been medicated for pain more effectively? And what about vulnerable people such as the aged, the poor, the disabled, or the uninsured being euthanized to relieve an economic burden? What would we think about a Brave New World in which the Hippocratic oath is turned upside down and physicians are cast in the role of killers, not healers?

The power of images, the experts tell us, far outlasts the impact of the spoken word. The images of Donald C.'s horrendous suffering, the cries of Donald C. howling in pain were useful to ground our ethical discussions in the harsh medical realities. But, I wondered, could the power of those images also subtlety distort the issues?

Then a student of mine, Babak Bina, had brought to my attention the plight of Mr. John Baker, a quadriplegic double amputee who suffered third degree burns on 80% of his body (which was, in fact, more severe than Donald C.'s injuries). John Baker had been attending Stony Brook's Adapted Aquatics Program run by Dr. Peter Angelo for the past ten years. However, John's Medicaid funding for transportation had been terminated due to a bureaucratic Catch-22.

As a result John had been confined to a nursing home where he was dying due to a lack of exercise. In the opinion of Dr. Charles Stewart of Stony Brook's School of Medicine, denying Mr. Baker his swimming would "undoubtedly contribute to the dampening of his remarkable human spirit which has kept him alive despite statistics showing a nearly 100% mortality in victims of burns this excessive."

After discussing the issue with my undergraduate teaching assistants, we decided we wanted to do something to help Mr. Baker. I called Dr. Peter Angelo and learned of all the dead ends he had run into in trying to help John. We invited John Baker to talk to my students about his experience as a burn victim and to share his personal perspective on the issue of euthanasia.

Getting Mr. Baker to campus turned out to be a major problem in itself. In the end, we were able to get a Stony Brook bus equipped with an hydraulic lift to transport Mr. Baker from the nursing home in Huntington to the university. In order not to overwhelm Mr. Baker, the student chose Jennifer McGinn to be the student who would go into the convalescent hospital to meet Mr. Baker before he was wheeled out to the school bus.

It was an exercise in futility to get campus news coverage of Mr. Baker's guest lecture. My teaching assistant Greg Lubicich suggested that we FAX a press release to Channel 12 News on the day John Baker was scheduled to speak. To our astonishment, Channel 12 picked up the story. It turned out that one of the news reporters, Connie Conway, had herself been involved in a serious life-threatening accidence, and perhaps there was no more pressing local news on that day.

As Channel 12 News filmed John being unloaded from the specially equipped Stony Brook bus, the reporter Connie Conway surprised John with the news that Medicaid had just decided to restore his funding!

The students saw Channel 12 filming the news story in front of their class in the Javits lecture center. Peter Angelo was interviewed and talked about how when he was about to give up, our class came along and took up the struggle. He spoke eloquently to my class about the value of love in encouraging each other to persevere in the face of obstacles.

When I announced to the students (many of whom had circulated petitions for Mr. Baker) that John's funding had been restored, my students clapped and cheered.

Then John Baker from his wheelchair shared his life experiences and insights with the class with refreshing honesty and humor. As I looked

Mr. Baker working on a jigsaw puzzle.

out among the sea of faces, I saw that some of my students had tears in their eyes. A week ago a student had said that society should do something; this week we realized that we are society.

Later that evening my students were invited to Dr. Angelo's Adapted Aquatics program to see John Baker demonstrate his famous 1 meter jump that had enabled him to win a gold medal for his participation in the 1985 International Games for the Disabled. My students also saw their peers teaching blind children and children crippled with Muscular Dystrophy to swim. We saw a community of joyful students whose lives had been radically transformed by their work with the disabled.

Throughout that eventful day it became very clear that John Baker's presence on the campus not only promoted the University's goal of enhancing its diversity, but also added to its humanity by giving students a life-changing educational opportunity to see a human face behind complex issues raised by medical ethics.

Perhaps what we need is not so much to articulate our current ethical thinking about euthanasia, but to listen to the testimonies of people like John Baker who can inspire us in revisioning our ethics and in cultivating caring communities in which the lives of the disabled enrich all of our lives.

"How much is one life worth?" one of my students casually asked. Then he started to worry that his "pro-life" opinions would force him to rethink his views about abortion. Socrates would have been delighted. We live our ethics. Perhaps unexamined ethics, like the unexamined life, is not worth living.

Baker's completed project.

Strategies for Creating

1. Aim at excellence

Socrates in the *Apology* goes about philosophizing in the firm belief that "wealth does not bring about excellence, but excellence brings about wealth and all other public and private blessings (30b)." In Homeric usage, the term is not gender specific and connotes the highest effectiveness of one's faculties and potentialities–the term applying to male heroes as well as to female heroines. The Greek word for excellence *arete* links the notions of "moral virtue" and "knowledge" as in the Platonic maxim "virtue is knowledge."

2. Creative questioning: What about x?

The word 'question' is derived from the Latin *quaerere* (to seek), which is the same root as the word for 'quest.' A creative life is a continued quest and good questions are useful guides. Some of the best questions are not questions asked to elicit a predetermined response, but questions that are open-ended invitations to wonder.

In asking creative questions it is also important to distinguish between constructive criticism and mere negativism which stifles creativity. Sometimes it is not a good idea to discuss your ideas with a negative person too early, instead it is better to let your ideas crystallize to some degree. In their companion book to the PBS program *The Creative Spirit*, Daniel Goleman, Paul Kaufman, and Michael Ray list a set of questions that illustrate the maxim that "there's no question too dumb to ask."[11]

- Bill Bowerman asked "What happens if I pour rubber into my waffle iron?" and was told that it was really a stupid idea but he went on to invent Nike shoes.

- Fred Smith posed this question in a paper for a business class at Yale: "Why can't there be a reliable overnight mail service?: Although he got a 'C' on his paper, he went on to found Federal Express.

- "Why can't we see in three dimensions what is inside a human body without cutting it open?" asked Godfrey Hounsfield who was told his idea was "impractical" but he went on to invent the Computed Axial Tomography (CAT) for which he was awarded the Nobel Prize in 1979.

- Masaru Ibuka was told "a recorder with no speaker and no recorder—are you crazy?" but went on to invent the SONY Walkman, which revolutionized the way music is distributed and listened to.

When some tells you that you've asked a "dumb question," perhaps you can take encouragement in Ralph Waldo Emerson's observation, "In every work of genius we recognize our own rejected thoughts."

3. *A far-sighted consistency*

Everyone knows that Emerson said that "a foolish consistency is the hobgoblin of little minds," which he said was "adored by little statesmen and philosophers and divines." There is a difference between a "foolish consistency" and a "far-sighted consistency" that keeps its eyes on the prize. The successful accomplishment of any project requires the perseverance to see the job through to its completion.

4. *Exploring other fields – fooling around is serious business*

Set time aside to explore and cultivate other fields or hobbies that stretch you mentally. "To err is human; to forgive, divine," wrote Alexander Pope (1688-1744) in his *Essay on Criticism* [1711], which is not an essay but a poem written with rhyming heroic couplets. Pope didn't coin the phrase "to err is human," which is the traditional English translation of the ancient Latin proverb. The word er, according to its Indo-European roots, meant "to move," which gave rise to the Latin verb errare, meaning to wander or to roam. The English word erratic describes a motion that is unpredictable and aimless, from which we get the word error. The Knight errant—including the prodigal Don Quixote who "reads the world in order to prove his books"—is on a quest, wandering on purpose and with the purpose of righting wrongs, and, more importantly, with the deep desire to experience something more of the adventure this life can be. To err, then, is to move into life, to wander, to be on a quest.

5. *Begin with an affinity for something – like falling in love*

Albert Einstein's fascination with physics began when he was just five, when he was ill in bed. His father brought him a present—a small compass. For hours, Einstein lay in bed, entranced by the needle that infallibly pointed the way north. When he was close to seventy, Einstein said, "This experience made a deep and lasting impression on me. Something deeply hidden had to be behind things."

Young ALBERT EINSTEIN

Many of the most profound transformational movements have been founded on love. The Suzuki method of teaching violin was birthed in the devastation of World War II when Shinichi Suzuki (1898-1998) noticed that it was natural for the children to speak their native language and reasoned that if taught with love they would also have the ability to become proficient on a musical instrument. His observations and writings are collected in *Where Love is Deep*.[12]

Dorothy Day was an American journalist, social activist, and the founder of the Catholic Worker movement. Tom Cornell, a former editor of the *Catholic Worker*, relates this story about Dorothy Day, that illustrates the kind of extravagance that belongs to any proper act of charity.

> One day a woman came in and donated a diamond ring to the Worker. We all wondered what Dorothy would do with it. She could have one of us take it down to a diamond exchange and sell it. It would certainly fetch a month's worth of beans. That afternoon, Dorothy gave the diamond ring to an old woman who lived alone and often came to us for meals. "That ring would have paid her rent for the better part of a year," someone protested. Dorothy replied that the woman had her dignity; she could sell it if she liked and spend the money for rent, a trip to the Bahamas, or keep the ring to admire.

DOROTHY DAY (1897-1980)

'Do you suppose God created diamonds only for the rich?'[13]

Dorothy Day's vocation was not only to call on us to reflect on the injustices of the world but also on the mysteries of love at work in our world.

6. *The logic of figuring out things on your own*

Richard P. Feynman (1918-1988), a self-described "curious character," was awarded the Nobel Prize in 1965 along with Sin-Itiro Tomonaga and Julian Schwinger for developing the theory of Q.E.D. or quantum electrodynamics. When he was 62, Richard Feynman reflected on a childhood experience that shaped his entire life. His father had a deep interest in science and liked to pose puzzles for his son to think about. Richard once asked his father why, when he pulls his red wagon forward, a ball rolls to the back.

> "That," he says, "nobody knows. The general principle is that things that are moving try to keep on moving, and things that are standing still tend to stand still, unless you push on them hard. And he says, "This tendency is called inertia, but nobody knows why it's true." Now that's a deep understanding.[14]

The message that the world is full of wondrous mysteries to be solved and the moral that "naming is not the same as knowing" stayed with Feynman all his life. Feynman figured things out for himself, from basic intuitions, and prided himself on being able to make the most complicated ideas in physics understandable to students. Feynman's lectures were masterpieces of exposition because Feynman celebrated the joy of figuring things out for himself and was able to communicate that joy to others. The following anecdote captures his spirit:

> Feynman was a truly great teacher. He prided himself on being able to devise ways to explain even the most profound ideas to beginning students. Once, I said to him, "Dick, explain to me, so that I can understand it, why spin one-half particles obey Fermi-Dirac statistics." Sizing up his audience perfectly, Feynman said, "I'll prepare a freshman lecture on it." But he came back a few days later to say, "I couldn't do it. I couldn't reduce it to the freshman level. That means we don't really understand it."[15]

RICHARD FEYNMAN

In his lecture *The Character of Physical Law*, Feynman talks about the role of imagination in science: "Our imagination is stretched to the utmost, not, as in fiction, to imagine things which are not really there, but just to comprehend those things which are there."[16]

7. *Expanding the spectrum*

Howard Gardner in *Frames of Mind: A Theory of Multiple Intelligences* [1983] proposed a theory of multiple intelligences. Rather than focusing on one kind of intelligence as defined and measured in the work of Alfred Binet and William Stern who coined the term "IQ" for intelligence quotient, Gardner wanted to acknowledge different kinds of intelligences. Here is his original list of seven kinds of intelligence: musical-rhythmic and harmonic, visual-spatial, verbal-linguistic, logical-mathematical, bodily-kinesthetic, interpersonal, and intrapersonal.

Over the years Gardner added to his list such intelligences as: naturalistic, existential, pedagogical and even computational. As with most writing on intelligence, Gardner's views were criticized for being *ad hoc* and arbitrary, for not being scientific or empirically testable, and for using the term "intelligence" too loosely as synonymous with "talents", "aptitudes", or even "personality".

We're not going to try to define "intelligence" or to demarcate science from pseudo-science (a topic of another module). Instead, we want to use Gardner's categories of intelligence as catalysts to expand our thinking about our own palette of critical thinking skills.

We began this module with asking you to take an inventory of your ideas about critical thinking. An inventory can be a catalyst for change: it can encourage you to enhance the skills you have and to explore new kinds of skills you wish to acquire. Rather than trying to find a paradigm case for each of Gardner's categories, let's play the game of trying to integrate the intelligences.

Instead of trying to isolate *musical intelligence*, think of how becoming musical, like Yo-yo Ma, requires not only dexterity, hand-eye coordination, and posture (*kinesthetic intelligence*) but involves performing with and for others (*interpersonal-intrapersonal intelligence*). Memorizing music not only improves *verbal-linguistic intelligence* but often can, especially in the music of Bach, facilitate the perception of musical patterns (*logical-mathematical intelligence*). The many venues in which Ma performed over his career required thinking about acoustics (*spatial-harmonic intelligences*). Moreover, his many musical workshops with children (*pedagogical intelligence*) and his many creative collaborations with other musicians, placed Yo-yo on a musical journey that forever changed his approach to living a musical life (*existential intelligence*).

It took me way beyond what I knew, into places of which I was totally scared, but as I became less frightened, I welcomed new ways of thinking and approaching something. It made me an infinitely richer person, and I think a better musician.
—YoYo Ma

8. *Networking with those in the know*

In almost any discipline or workplace there are individuals who have become repositories of wisdom. One good way to get to know your way around is to network with these experts. Why do Nobel prize winners tend to reproduce successful researchers–even though everyone has access to their publications? One answer is that when you are in the presence of a creative individual you pick up lessons about their habits of thought, how they go about solving problems, how they work–you pick up tacit knowledge that is not contained in their publications.

9. *Collaborative excellence*

Explore different models of winning. The competitive model of success has dominated our thinking about earning and grading—the winners are honored with the spoils in a zero-sum game. The competitive or victory model of excellence promotes individualistic thinking about excellence. Collaborative models of success emphasize teamwork—thinking about how win-win situations can arise out of non-zero-sum games. The cooperative model of excellence promotes relational thinking about excellence.

This is just a partial list of strategies for creating lives more attuned to the challenges and joys of problem solving. As you become more adept at creating lives, you will come up with others.

The American poet Robert Frost (1874-1963) wrote a delightful essay *The Figure a Poem Makes* [1939] in which he expresses his personal views about poems. He also happens to capture something which I have felt about how one learns logic in retrospect–how things that seem difficult the first time through can be seen as familiar friends when one looks back, retrospectively, over what one has learned:

> I tell how there may be a better wildness of logic than of inconsequence. But the logic is backward, in retrospect, after the act. It must be more felt than seen ahead like prophecy.
>
> The figure a poem makes. It begins in delight and ends in wisdom.... For me the initial delight is in the surprise of remembering something I didn't know I knew.... There is a glad recognition of the long lost and the rest follows. Step by step the wonder of the unexpected supply keeps growing.

The Apple of the Mind's Eye

We'll conclude this module on problem solving with an image that recalls the initial story from which we began. It is a metaphor that takes the forms of another story about a child's encounter with an apple. It is from D. N. Perkins's beautiful book, *The Mind's Best Work*:

> A couple of years ago, my oldest son accosted me with news from kindergarten as I came home from the office. He had learned something about apples and wanted to demonstrate. Out of the drawer came a knife, one of those he was not supposed to handle, and out of the refrigerator a McIntosh. "Dad," he said, "Let me show you what's inside an apple."
> "I know what's inside an apple," I said, riding for a fall.
> "C'mon, just let me show you."
> "Listen, I've cut open lots of apples. Why ruin an apple just to show me something I already know?"
> "Just take a look."
>
> Ungracefully, I gave in. He cut the apple in half, the wrong way. We all know the right way to cut apples. One starts at the stem and slices through to the dimple on the bottom. However, he turned the apple on its side, sliced the apple in half perpendicular to the stem, and displayed the result.
> "See Dad. There's a star inside."[17]

Whoever first sliced an apple the wrong way may well have had a good reason to do so, curiosity being one good reason. Or it might have been one of those fruitful—I choose the word carefully—mistakes all of us make sometimes. ...The knowledge of it traveled from unknown origins to my son's kindergarten class and so to me and you.
—D.N. Perkins

This module began and ended with an apple. We began with Einstein's admonition about imposing our wills upon the chaos of nature and ending with a parable about a child's joy at discovering a star in an apple sliced the wrong way.

Summary of Concepts

Puzzles provide a CONTROLLED way of studying various aspects of problem solving but they are limited by being artificial. However, this approach is artificial in the same way as a scientific experiment is designed to control for a single variable is artificial.

One fascinating aspect of problem solving is how we use our MEMORIES from the past to solve problems in the future. Memory tricks can be used to extend our memories to recall beyond George Miller's magic number that found human beings can typically remember about 5 + 2 bits of unrelated information.

The PEG WORD system connects items in a laundry list of items to images pegged to a rhyming scheme. The classical method of the MEMORY PALACE connects the sequence and key points of a speech to walking through the rooms of a familiar house or to traveling a well-known path or journey.

As one grows older there is less need to memorize laundry lists of items and more of a need to organize one's knowledge in CONCEPTUAL HIERARCHIES (or "ontologies" as they are now called in designing human interfaces on the internet.)

Another powerful problem solving method is to EMBED a problem in a richer, and well known problem space, e.g., embedding Euclidean geometry in Cartesian analytic geometry. Once this is done ISOMORPHISMS, or structural similarities, between seemingly difference problems become more obviously apparent allowing for a more general, and logically unified, approach to problem solving.

COGNITIVE EMOTIONS, e.g., in addition to cognitive intellectual skills, can be a way of knowing. As Einstein famously remarked, "Imagination is more important than knowledge" by which he didn't mean that knowledge was unimportant, but that a love for truth and simplicity, and repugnance towards errors in logic or fact, can themselves point to new ways of knowing.

Problem solving, in addition to being a set of skills or heuristics, can be an APPROACH TO LIVING. Instead of ignoring problems, we can use them as occasions for expressing our creativity.

Exercises

Group I

1. Memory Games

Make up your own memory system for playing Hot and Ace without looking (playing "blindfold"). For example, you could extend the previous peg-word system by tagging the pairs of words and numbers with their positions on a Tic-Tac-Toe board. Or perhaps you can figure out a way of visualizing the magic square by generating the sequence of numbers using a knight's move from chess. Or starting from scratch, you may find a coding easier to use when the memory system begins with positions (labeled alphabetically) and then tagging them with content consisting of pairings of words and numbers. Enjoy inventing your own memory system –be it pegs or palaces or something poetic. Here's the relevant matrix of information for coding:

A	B	C
Brim-2	Hot-9	Wasp-4
D	E	F
Woes-7	Hear-5	Tied-3
G	H	I
Tank-6	Ship-1	Form-8

2. JAM

Here's a third game invented by Dutch psychologist John A. Michon.[18] The game is called 'JAM' since those are the initials of the inventor. To the right is a map of cities (the dots) connected by roads (the arrows) Two players take turns coloring a length of road. The first player to color three roads that pass through the same city is the winner.

Show that this game is isomorphic to HOT and ACE.

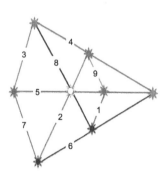

3. *Frames of Mind and Intelligence Reframed*

(A) Think of examples to illustrate each of Howard Gardner's examples of different kinds of intelligences.

(B) Can you think of any other kinds of intelligences not discussed by Gardner? List them, give clear examples of each, and then suggest ways to develop such intelligences.

(C) In his retrospective view of his work on "multiple intelligences" Howard Gardner argues that emotional intelligence and moral intelligence are not types of intelligences. What is Gardner's argument for these claims?

Group II

4. *Verbatim*

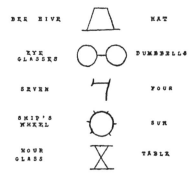

Design an experiment to determine the effects of verbal labels on memory. Carmichael, Hogan, and Walter for example, presented the ambiguous figures in the center column with different verbal labels. The labels they obtained with their respective figures are shown in the first and third columns.[19]

5. *Seriously Wired*

There are three switches A, B and C and a light on the top of a box, which contains the secret wiring inside. The magician asks a spectator to choose one of the three switches.

1. The spectator flips the chosen switch and the light goes on and then again and the light goes off.

The magician flips the other two switches to show they do not turn the light on, and then she challenges the spectator to find the switch that controls the light.

2. The spectator chooses his original switch, but when he flips the switch the light doesn't turn on.

The magician then gives the spectator another choice.

3. The spectator choses one of the remaining two switches, but it too doesn't turn the light on.

The magician then chooses the remaining switch, which turns on the light and off.

How can you discover the secret wiring which does the trick? [Hint: be serious.]

6. *Loop of the Rings*

Sometimes when we become adults, we stop puzzling about everyday phenomenon. Leonardo di Vinci once wrote in a letter to a friend, "You and I never cease to stand like curious children before the treat mystery into which we were born." This simple trick requires a loop of chain about 3 feet in length and a several metal rings 2 inches in diameter and a larger ring 2.5 inches in diameter.

Drape the chain over the middle three fingers of one hand (say the right) or between the thumb and forefinger as in the diagram. Place the loop of chain through the ring and hold the ring with the thumb and middle finger of the right hand. Now simply let go of the ring and let it slide down the chain with you other hand in place ready to catch the falling ring. Keep doing until something rather surprising happens.

Try to explain to yourself what is happening. You might try experimenting different variations—for example, my friend and I succeeded in looping up to five 2-inch rings at once and even was able to loop a small ring whose diameter is only 1.5 inches. (The three concentric rings can make a charming necklace.) The point of this exercise is not to memorize some explanation from physics but to recover a playful child-like wonder that is relentlessly and randomly curious.

Group III

7. *Fractal Tic-Tac-Toe*

Played on a Sudoku board—in which the nine cells of the standard tic-tac-toe board themselves contain a smaller tic-tac-toe board. Each cell on a small board corresponds to the cell on the big board, e.g., the middle cell in the top row of any small board corresponds to the middle cell in the top row of the large board.

1. Players take turns as in regular tic-tac-toe.
2. Each play on a small board determines where the next player must play on the big board.
3. Once a tic-tac-toe game is won on a small board in a cell on the large board that cell is replaced with a large X or O, depending on which side won.
4. When a move in the required square is not possible, then the player has a free choice to play anywhere.
5. The player who gets three in the row on the large board wins.

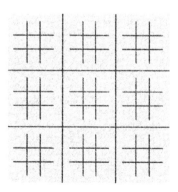

The object of this exercise is to create a fun version of 3-dimensional (or even 4-dimensional) fractal tic-tac-toe. Once one starts to get the sense that a loss seems inevitable, playing on loses much of its charm. To remedy this, add an element of chance so that even a losing position can still be won.

8. *Patterns & Paradigms*

Suppose you are presented with a sequence of shapes, which you are to organize into an over-arching pattern or paradigm. This paradigm allows you to make an orderly pattern in the stream of shapes. Let's say these shapes are, metaphorically, are the shape of "facts" that you must fit into a "theory".

The first singular fact you are given is a simple aquamarine square. Naturally, enough you hypothesize: the facts of reality form a square. Next, you receive a pair of facts—two small right triangles. You can't form all three shapes into a square, but you can, building upon your original paradigm, you can arrange the shapes into a double square or rectangle. Thirdly, you receive the pair of large right triangles—one rose and one purple. Again, you find it fairly easy to accommodate this pair of facts into your rectangular paradigm.

The next pair of shapes poses an anomaly for your pattern, however. How can you fit the single medium sized right triangle and the parallelogram into your pattern? All of this causes you to question your paradigm and to search for a new one.

Fool around and come up with an new organizing paradigm that accommodates all the facts.

There are several possibilities.
One preserves the square hypothesis.
Another introduces a totally new paradigm based on the parallelogram.
A third combines the two.

"... novelty ordinarily emerges only for the [scientist who], know[s] with precision what [to]... expect, [and] is able to recognize that something has gone wrong."
—THOMAS S. KUHN (1922-1996)
The Structure of Scientific Revolutions

In the early 1960s Thomas Kuhn published *The Structure of Scientific Revolutions*, an influential book about the history and philosophy of science. Kuhn portrayed the history of science as proceeding in stages that alternate between "normal" and "revolutionary" science. During the stage of "normal science" scientists make progress by "puzzle solving"—fitting facts to confirm the currently accepted explanatory schemas, which Kuhn called paradigms. During these "puzzle solving" periods, science progresses by accumulation of facts.

However, certain facts, which Kuhn called anomalies, resist fitting into the reigning paradigm, exerting pressure for a revolution in science. Typically, the heroes of science are the architects of these "scientific revolutions" where a new paradigm emerges that changes the fundamental ways that science has proceeded. During these paradigm shifts, the facts that accumulated under the old paradigm may have to be re-conceptualized, and some simply ignored as "incommensurate" with the new paradigm.

This puzzle was designed to give you a sense of how such a paradigm shift might be experienced. Kuhn's model of the history and philosophy of science will be critiqued in a future module of *Thinking Matters* dealing with Scientific Reasoning, Probability and Decision Making.

9. *Eleusian Rule-Following*

What numbers come next the following sequences?

(A) 1, 8, 15, 22, 29, 36, ___.
(B) 1, 2, 4, 8, 16, 32, ___.
(C) 1, 2, 5, 10, 17, 26, ___.
(D) 1, 8, 27, 64, 125, ___.
(E) 1, 2, 3, 5, 8, 13, ___.

Perhaps you remember these problems from "intelligence tests" or elementary school arithmetic. Usually you are asked to explain your answer by citing a rule involving such operations as addition, multiplication, exponentiation or the Fibonacci rule. Before moving on to the point of this exercise, you're probably wondering how you did. Here's a coding (in reverse) of the *intended* answers: 126, 127, 346, 340.

Here's a model for finding solutions to number sequencing problems. Fill in the missing numbers:

```
    1   9   19   33   ?   83
     +▽  +▽  +▽   +▽  +▽
      8  10   14  (_) (_)
       +▽ +▽  +▽  +▽
        2   4  (_) (_)
         ×▽ ×▽ ×▽
          2   2  (_)
```

If you're interested in more gimmicks like this one to improve your performance on intelligence tests, see *Thinking Better* by David Lewis and James Greene. What's the difference between a gimmick and a method that generalizes? A method that generalizes is a gimmick you use at least twice.

Ludwig Wittgenstein advanced a skeptical argument about rule-following which was posthumously published *Philosophical Investigations* [1953]. The philosopher Saul Kripke in his commentary *Wittgenstein on Rules and Private Language* [1982] called this argument "the most radical and original skeptical problem that philosophy has seen to date." (Due to the ambiguity of Wittgenstein's enigmatic remarks and to the originality Kripke's insights, this book is said to

> *This was our paradox: no course of action could be determined by a rule, because any course of action can be made out to accord with the rule.*
> —Wittgenstein(1889-1951)
> *Philosophical Investigations* [§201a]

be an insightful a commentary on the work of the philosopher "Kripkenstein." The so-called "adoption problem" for logic tracing back to Lewis Carroll's parable "What the Tortoise Said to Achilles" [1895] is explored in the module *Critical Thinking as Logic and Deduction*.)

To illustrate this paradox, what's the next number in the sequence,

$$(K)\ 1, 4, 9, \_\_?$$

It is natural to suppose that the "right answer" must be 25. If so, then what's the next number in the sequence? On the other hand, instead of 25 suppose the next number is 27. Can you find a rule that generates this and the next number in that sequence? Or suppose the next number is between these two, namely, 26. Can you find a rule that generates neither 36 nor 40, but, say, 33? If fact, given any arbitrary number like 57, we can generate a mathematical formula that generates it as the next number in the sequence, e.g., consider the "roots" or solutions to the following polynomial:

$$(x-1)(x-4)(x-9)(x-57) = 0$$

One moral to draw from all this is that in mathematics, induction means something quite different than it does in the empirical sciences. These ideas are developed in the module on *Scientific Reasoning and Probability*, which discusses Hume's Problem of Induction, Hempel's Raven Paradox, and Goodman's Grue Paradox as well as a model of learning from experiences known as Bayesian reasoning.

10. *Illeism*

Excessive self-rumination can be paralyzing. Ask Hamlet. There is an ancient rhetorical device for known as illeism in which one talks about oneself in the third person. It was deployed by Julius Caesar in his *Commentaries on the Gallic War* to create a sense of objective achievement, and the Royal "we" has been used by Kings and Queens, or pompous politicians, to create a sense of impersonal power. It has also been used for various literary effects—to hide the identity of the narrator/culprit in detective literature, to suggest a lack of self-awareness in artificial life forms, such as robots or androids, in science fiction, or to suggest a lack of personal worth in a master-slave relation as in the speech of Dobby the house elf in *Harry Potter* or Gollum in *The Hobbit*. Some psychologists speculate that this rhetorical device can be used to overcome excessive rumination.[20]

Come up with some heuristics for deciding when to use illeism or its opposite to "get in touch with one's feelings" or when to deploy objective argument or personal narrative to make a persuasive appeal.

11. *The Color of Logic*

The idea for some of the puzzles below came from a chance conversation at a wedding reception with a colleague. As we waited to be seated, this colleague, a linguist, was advancing the thesis that logic was based on the color lexicon and the physiology of color perception.

From a logician's perspective, this seems to reverse the order of things. The color wheel's structure is explained by logic, not logic by the structure of the color wheel. Lacking the details, I sketched out my own mapping between color and logic.

It turned out that our mappings were different, but isomorphic. He (and a colleague) had mapped the additive color wheel for light onto the Blanché Hexagon for propositional logic; I had mapped the subtractive color wheel for paints onto the Aristotelian Square of Opposition for categorical logic. Logic is more fundamental: it gives the common structure of both (contingent) color wheels.

Here's the chance for you to make these discoveries for yourselves.

First, a little background. Philosophers have raised some intriguing questions about colors.

- Do colors exist in the physical world or do they existence subjectively in the mind?

When I talk to people at parties and they learn I'm a philosopher, they usually remember the famous question from Philosophy 101: "If a tree falls in the forest and there's no one there to hear it, does it make a sound?" They rarely remember the point of the question. Rationalist philosophers like René Descartes and empiricist philosophers like John Locke distinguished between primary properties (which are supposed to exist objectively in the object) from secondary properties (which exist subjectively because of how the object causes certain perceptions to occur in observers). Colors are thought to be of this secondary kind.

- The philosopher Frank Jackson [1982] posed a puzzle known as the "Mary's Room" thought experiment. Initially Jackson deployed this puzzle to try to refute a view known as *physicalism*, but he later changed his mind (but not his brain!) about whether the thought experiment refuted physicalism.

The thought experiment, nevertheless, is a good one. Mary is a color scientist, who is confined to a room, but who has a computer by which she can access all the information about colors–except that all her information is coded in uncolored bits and her computer screen only displays in shades of gray. Can you know everything there is to know about color by coding all our knowledge–e.g., cyan has wavelength = 490-520 nm, frequency = 610-575 THz, hex triplet = ##00FFF, sRGBN = (0, 255, 255), CMYK = (0, 255, 255)—into colorless bits of information? Then one day Mary escapes her room and walks into the outside world where she sees colors—e.g., the qualia or sensation of red—for the first time.

Does Mary now know something she didn't know before–that is, does her experience of the qualia of color sensation give Mary knowledge about color goes beyond the information she knew before?

- A Phenomenological Experiment: Stare at the red square on the left for 20 seconds and then shift your vision to the gray square on the right. What color do you see?

Aristotelian color theorists had argued, on the basis of their *a priori* theories, that the afterimage of a red object must be green. There's a problem with this theory: the afterimage isn't green. What color is the afterimage? Find out about how the eyes' photoreceptors, rods and cones, function in color perception and how this phenomenon may be explained.

These are three fascinating questions raised in the philosophy of perception. Now for a little history.

The idea that light is composed of colors is associated with Isaac Newton (1642 — 1729/27, the story why there are two years for Newton's death is an interesting one, which we shall forgo). In his 18th century scientific treatise of the Enlightenment, *Optiks* [1704], Isaac Newton famously demonstrated by means of a prism that white light could be refracted into prismatic colors. It is a common misconception that Newton believed that light was composed of colors or that colors were in light. In fact, Newton believed that light has the disposition to trigger colors in one's visual system. Newton did not explain how these colors are triggered. Although color diagrams were not new to him, Newton systematized the spectrum of colors into a wheel in which pairs of complementary colors of light "cancel each other out" to produce white light.

You may have been taught that when a white beam of light is projected through a prism, it is refracted into a rainbow of seven colors–Red, Orange, Yellow, Green, Blue, Indigo and Violet (remembered by ROY G BIV). If you've ever had trouble distinguishing Indigo and Violet, you're not alone. Even human beings with good eyesight can have difficulty distinguishing indigo from dark blue. Violet, which is easier to distinguish, clearly belongs to the short-wavelength colors sensed by the blue or S-cones.

It turns out that Newton's motivation for inventing the myth of the seven color rainbow had to do with the fact he wanted to map the seven colors of the rainbow onto the seven tones of the musical scale. This was his reason for distinguishing Violet and Indigo. The first recorded use of "indigo" as an English color name was in 1289. It turns out that it is more natural for us to divide the spectrum into six, rather than seven, bands of color. Six is a better fit for the logic of mixing three primary colors.

The color wheel for painters was discussed by Johann Wolfgang von Goethe (1749-1832) in his *Theory of Colors* [1810]. When complementary paints are mixed they produce brown. The conflict between Newton's scientific theory of color of light and Goethe's theory of the color paint is classic conflict of cultures. It turns out both theories are correct. There are two systems for specifying colors—RGB for web design and CMYK for print design.

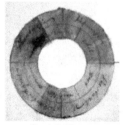

Goethe's *Theory...*

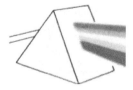

Newton's *Optiks*

"During the day, owing to the yellowish hue of the snow, shadows tending to violet had been observable: these might now be pronounced decidedly blue as the illuminated parts exhibited a yellow deepening to orange. But as the sun at last was about to set and its rays greatly mitigated by the thicker vapors began to diffuse a most beautiful red color over the whole scene around me, the shadow color changed to a green in lightness to be compared to a sea-green in beauty the green of an emerald."

—Goethe

192 THINKING MATTERS: CRITICAL THINKING AS CREATIVE PROBLEM SOLVING

Color Wheel for Paint
(Subtractive)

Color Wheel for Light
(Additive)

In the painter's color wheel, there are three primary colors—red, blue and yellow. Mixing then in pairs produces three secondary colors—purple, green and orange. Mixing adjacent primary and secondary colors produces the tertiary colors and a wheel of 12 colors. The color wheel for light, begins with three primary colors—cyan, magenta and yellow—and mixing these colors of light in pairs produces—red, green and blue, etc.

Now for some puzzles:

(A) Find a mapping of colors onto the Aristotle Square of Opposition that produces such color theorems as:

A	universal affirmative	"All ○ are □"
E	universal negative	"No ○ is □"
I	existential affirmative	"Some ○ are □"
O	existential negative	"Some ○ are not □"

ARISTOTELIAN SQUARE

1. ● ∩ ● ⇒ ●
2. ● ∩ ● ⇒ ●
3. ● ∩ ● ⇒ ●
4. ● ∪ ● ⇒
5. ● ∩ ● = ●   ● ∪ ● = ○
6. ● ∩ ● = ●   ● ∪ ● = ○

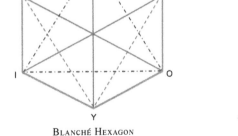

BLANCHÉ HEXAGON

(B) Find a way of labelling the nodes of Blanché Hexagon with the following propositions:

(P ∧ Q)	and
(P ∨ Q)	or (in the inclusive sense)
(P ∨ Q)	or (in the exclusive sense)
(P ↔ Q)	if and only if
(P ↓ Q)	neither nor
(P | Q)	not both

Then label and 'colorize" the hexagon satisfying the following desiderata:

i. The propositions should be labeled A, E, I, O, U, and Y.
ii. The contradictory pairs (A, O), (E, I), (U, Y) occupy

opposite corners of the hexagon and be colored in complementary colors.

  iii. The implications should make sense in terms of primary and secondary colors.

    A implies both I and U.
    E implies both O and U.
    Y implies both I and O.

(D) *"Neither, nor"* (NOR) and *"Not both"* (NAND) are the two, and the only two, logical connectives that are "primary" in the sense that each is adequate to express all the other connectives. Does this have any interpretation in terms of color?

(E) Can you discover (and explain) the isomorphism between the Aristotelian Square and the Blanché Hexagon?

(F) Can you think of a logical interpretation that makes use of the primary, secondary, and tertiary colors?

Preview

The famous logic puzzle known as the "Fork in the Road" (or, under various incarnations, "Angels and Demons," or as a puzzle in a popular logic textbook about which I used to get calls from logic teachers on how to solve) can serve as a window for the next module in that it raises intriguing questions about the basic conceptions of logic and deduction. The puzzle concerns a logician vacationing on a South Seas island inhabited by the two proverbial tribes—truth-tellers, who always tell the truth, and liars who always lie.

> A logician, whom we 'll call Alfred, comes to a fork in the road, one of which leads to the village and the other which leads to the cannibals. At the fork in the road, there is an inhabitant of the island, but Alfred doesn't know whether the inhabitant is a truth-teller or liar. What question can the Alfred ask so that from the native's reply, he'll be able to know which road leads to the village.

Try to figure out a logical answer to this puzzle.

What is logic? Gottlob Frege (1848 — 1925) is regarded as the inventor or discoverer of modern symbolic logic because of his groundbreaking publication known as the *Begrifftsschrift* [1879] (literally, "conceptual writing"). Frege used this logic to try to prove that all laws of arithmetic could be founded on pure logic. In the end, his life's dream was shattered by Bertrand Russell in 1901 by the discovery of what has become known as 'Russell's Paradox'. Although he considered his life's work a failure, Frege continued to write about logic. In an essay "The Thought: A Logical Inquiry" [1918-19], Frege wrote:

> The word 'true' indicates the aim of logic as does 'beautiful' that of aesthetics or 'good' that of ethics. All sciences have truth as their goal; but logic is also concerned with it in a quite different way from this. It has much the same relation to truth as physics has to weight or heat. To discover truths is the task of all sciences; it falls to logic to discern the laws of truth.

According to one intuitive conception, truth is a *correspondence* between one's thoughts and reality. We can diagram this relationship as follows:

Correspondence

Reality or the Actual World ←------------- Thoughts

This situation is obviously too vague to make progress, so the logician introduces suitable abstractions for each of the elements of this relation. Instead of thoughts, the logician introduces a *formal language*, which can be specified by giving an alphabet of symbols and a grammar for forming symbolic sentences from that alphabet. The reason for specifying a *formal* language is that the logical features of deductive thought that we wish to model depends on the form, rather than the *content* of the language.

Another element of this relationship is reality or the "actual world." Logic doesn't deal with truths that merely happen to be true about the actual world. The laws of truth must, in the colorful expression of the philosopher, mathematician and lawyer, Gottfried Leibniz (1646-1716), be true "in all possible worlds." So we abstract from "reality" or the "actual world" to the notion of a possible world. The idea of analyzing the truth of a sentence in all possible worlds is captured by the notion of a truth table, to which you have already been introduced. In fact, the notion of a "correspondence" can be rigorously modelled by a definition of truth in terms of those truth tables. At we shall see that the definition of truth in a possible world is the key to defining logical truth.

Correspondence

Reality or the Actual World ←------------- Thoughts

⇓ ⇓ ⇓

Possible Worlds   Definition of Truth   Formal Language
(Rows of Truth Tables)

All the ideas we obtain in this way by abstracting from the intuitive picture can be formally and rigorously formulated—and we shall do so in the module, *Critical Thinking as Logic and Deduction*. The intuitive idea is that logic is not concerned with what happens to be true in one world, but in the *laws* of truth: the laws of truth that are true in all possible worlds.

Keeping in mind this picture of what logic is, let's return to the truth-teller and liar puzzle and discover a solution which holds in "all logically possible worlds."

Alfred, being a logician, pulls out his logical calculator. Since he wants to ask a question, the answer to which will tell him whether to take the left or the right fork in the road. This can be modelled with a "yes" and "no" (or true and false) question. Alfred specifies his problem in a way such that to find a yes-no question X, the native answers as follows:

Question $X = \begin{cases} \text{"yes" if the left road leads to the village} \\ \text{"no" if the left road does not lead to the village} \end{cases}$

Calling up his truth-table app, Alfred chooses two statements from which to construct his question:

Dictionary:
P: You (the native) are a truth-teller
Q: The left road leads to the village

Since it is given that the native is either a truth-teller or liar and that one of the two roads at the fork leads to the village, then the negation of P, ~P, means the native is a liar and the negation of Q, ~Q, means that the right fork in the road leads to the village. Now our question is precisely specified.

The truth table app gives Alfred the following chart, which he begins to fill out. The answer Alfred wants is "yes" (or true) if the left road leads to the village and "no" (or false) if the right road leads to the village. So the "yes" and "no" answers wanted are to match the column under Q:

P	Q	Answer wanted	Question X to be asked
T	T	*yes*	
T	F	*no*	
F	T	*yes*	
F	T	*no*	

Alfred can now work backwards to figure out what the actual truth values for the question X should be. In the first two rows of the truth table, P is true, which means the native is a truth-teller. Since a truth-teller always tells the truth, then the correct answer to question X matches the values for Q.

The tricky part comes into play when the native is a liar, and so reverses the correct answer to question X. When P is false (in rows 3 and 4), the native is a liar and so will reverse the correct answer to the question. Therefore, the truth values for *X* in these rows must be the opposite of the answers wanted.

P	Q	Answer wanted	Question X to be asked
T	T	yes	T
T	F	no	F
F	T	yes	F
F	T	no	T

We now have precisely defined questions. Find a truth-functional statement composed of P and Q which has the truth table required for question *X*.

A simple answer is the *biconditional* P ↔ Q, i.e., Alfred asks the native, "Is it the case that you are a truth-teller *if and only if* the left road leads to the village?"

We can verify that this solution works by checking to see if it gives the right result for every logically possible world. There are four possible worlds in the diagram below representing all the logically possible worlds. The T or F at the fork in the road represents whether the native is a truth-teller or liar, the house at the end of the road represents the location of the village, and the numbers of the quadrants correspond to the four rows of the truth table. The question asked is represented by "(P ↔ Q)?", the truth-values under the parts of the sentence calculates the correct answer to the question, and the answer "yes" or "no" is the native answer, depending on whether he is a truth-teller or liar.

We verify that the first two possible worlds give a correct solution to Alfred's question.

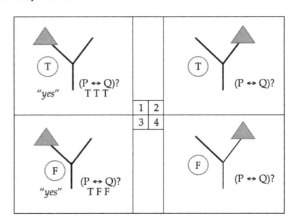

- In the first possible world (quadrant 1), the native is a truth-teller and the left road leads to the village. Therefore, the correct answer to the question (P ↔ Q) is true and so the native, being a truth-teller answers 'yes" and Alfred chooses the left road and ends up at the village.
- In the third possible world (quadrant 3), the native is liar and the left road leads to the village. Therefore, the correct answer to the question (P ↔ Q) is false and so the native, being a liar, reverses the correct answer and 'yes" and Alfred again choose the left road and ends up at the village.

Verify for yourself that the solution works for the possible worlds 2 and 4. We have a logical solution to our puzzle, one that is guaranteed to work for all logically possible worlds.

When Martin Gardner originally posed the puzzle, he received a delightful letter from a pair of his readers exposing some deeper logical conundrums. I quote from the letter:[1]

> It is sad commentary on the rise of logic that it leads the decay in the art of lying. Even among liars, the life of reason seems to be gaining ground over the better life.... If we accept the proposed solution, we must believe that liars can always be made the dupes of their own principles, a situation, indeed, which is bound to arise whenever lying takes the form of slavish adherence to arbitrary rules.

A popular solution to the puzzle involves a *counterfactual* conditional: "If you were to ask you if left road leads to the village, would you say 'yes'?" The authors of the letter point out the solution's problem:

> [E]xpecting [the native] to have interpreted this question as counterfactual conditional in meaning as well as form, presupposes a certain precocity on the part of the native. ...If [Alfred] asks the question casually the native is almost certain to mistake the odd phraseology for some civility of manner taught in Western democracies, and answer as if the question were simply, "Does this road lead to the village?" On the other hand, if he fixes him with a glittering eye in order to emphasize the logical intent of the question, he also reveals its purpose, arousing the native's suspicion that he is being tricked. The native, if he is worthy of the name of liar, will pursue a method of counter-trickery....

As we shall see in Module II of *Thinking Matters, Critical Thinking as Logic & Deduction*, counterfactuals have intriguing logical properties, which are not adequately modelled by the truth-functional conditional. These issues concern the "paradoxes of material implication."

A variant of this puzzle concerns a choir composed of a mixture of Angels (who always tells the truth) and Demons (who always lie). Beside the choir are two doors one labelled "*Oom*" and "*Doom*"–one leads to Heaven and the other to Hell. You know that either *oom* means *yes* and *doom* means *no*, or that *oom* means *no* and *doom* means *yes*, but you don't know which. This is important because the door to Heaven is labelled with the word meaning *yes* and the door to Hell is labelled with the word meaning *no*. You get to ask a single question to which the choir will answer in unison–a potential cacophony of *ooms* and *dooms*.

As a matter of confessional courtesy, Angels and Demons won't answer questions about whether they or a another choir member is or is not an Angel or a Demon. I forgot to mention the dress code for the choir: the Angels dress in white and all the Demons in black, but this may not be that useful because they are all invisible to you. Suppose you want to know which door lead to Heaven. Can you find a single counterfactual question that will do the trick?

The authors of the letter to Gardner go on to distinguish different several types of liars: (1) *simple liars*, who simply reverse the truth value of a complex sentence at the end of their calculation; (2) *honest liars*, who impartially lie to themselves before calculating the truth value of their response (raising the issue of *logical duality*, a topic taken up in the next module), and finally (3) *artistic liars*, which they explain in the final paragraph of their entertaining letter.

> Clearly the essential feature of [logician's] strategy must be its psychological soundness. Such strategy is admissible since it is even more effective against the honest and the simple liar than against the more refractory artistic liar... .

If you wish to know the most general solution posed by these authors, I suggest you check out their letter to Martin Gardner, reprinted in the (First) *Scientific American book of Mathematical Puzzles and Diversion* (p. 32), or to proceed to Module II in which we'll provide a method that enables you to deduce immediately why our solution to the puzzle works by knowing a couple of theorems of propositional logic.

The solution to this puzzle is similar to the following one, in which you agree to answer two "yes/no" questions, the first of which is:

If I were to ask you, "Will you go on to read the next module of *Thinking Matters* would you answer *that* question the same way as *this* one?

Epilogue

There is a reason for dividing *Thinking Matters* into modules. Each module takes as its subject matter a particular domain of critical thinking with its characteristic thinking skills. But the domains are not distinct. One of the joys of critical thinking is making a creative leap that synthesizes different domains and synergizes your thinking skills. Let's conclude with one such example.

The logical connective Xor (in symbols, $\veebar$) is the *exclusive 'or'*: given two propositions, P and Q, (P $\veebar$ Q) is true iff exactly one of P and Q is true. You have already verified the first two facts about Xor to which we add a third:

i. (P $\veebar$ Q) is equivalent to (P $\vee$ Q) $\wedge$ $\sim$(P $\wedge$ Q), literally, "P or Q but not both";
ii. it is also equivalent to the *negation of a biconditional*, i.e., any of $\sim$(P $\leftrightarrow$ Q), (P $\leftrightarrow$ $\sim$Q), or ($\sim$P $\leftrightarrow$ Q);
iii. but the binary connective does not generalize in the usual way, i.e., (P $\veebar$ Q $\veebar$ R) $\neq$ ((P $\veebar$ Q) $\veebar$ R).

Xor is generalized in programming languages such as *Python* as a bitwise operation on an array of n-ary binary strings that yields 1 (true) if there is an *odd number* of 1's in that binary column and 0 (false) otherwise. This generalization of Xor calculates *parity*, the evenness or oddness of a sequence of 1s or 0s. Parity, you might recall, was the key to Euler's solution to the Königsberg Bridge puzzle.

The game of Nim can be analyzed mathematically using binary numbers. Write the number of counters in each row in binary, arranging the binary numbers so that their digits are aligned in vertical columns starting with the 1's place. A *winning position* is reached when the *parity* of each column of the binary string is even, i.e., the sum of the 1s in each binary column is even. This non-carrying binary sum is Xor. It's your turn to move in the following game of Nim. Can you find the winning move?

> *Our total reality and total experience are beautiful and meaningful–this is also a Leibnizian thought. We should judge reality by the little which we truly know of it. Since that part which conceptually we know fully turns out to be so beautiful, the real world of which we know so little should also be beautiful.*
>
> —Kurt Gödel
> confiding to Hao Wang[1]

♠ ♠ ♠ ♠ ♠	5	0101
♥ ♥ ♥ ♥ ♥ ♥ ♥	7	0111
♣ ♣ ♣ ♣ ♣ ♣ ♣ ♣ ♣	9	1001
	Xor	1011

Parity is a ubiquitous idea. It turns out that XOR is a fundamental idea used by Hamming codes to check, and even locate with a high probability, the data that have been corrupted while sending or storing a message. (I recommend a delightful video by Grant Sanderson on the 3blue1brown.com *YouTube* channel dealing with Hamming Codes.)

I would like to end this module with a Magical XOR-ism, a card trick you can perform blindfolded. Ask a spectator to hand you four randomly chosen cards. As you hold the cards facing the spectator, ask him to choose a card as you move your finger over the tops of the cards. Then follow this algorithm:

1. Skip the card that's adjacent to the chosen card so you can flip over the card next to it. (The back of the flipped card now faces the spectator)

Then ask the spectator to instruct you on how to shuffle the deck.

2. Do you want me to cut the deck after the first, second, or third card? Cut the deck.
3. Then flip over the top pair of cards together and place them on either the top or bottom of the deck.
4. Then alternately deal the cards into two piles.
5. Ask the spectators to choose one of the piles, then take that pile, flip it over, and place it on top of the other pile.
6. Fan the cards. The chosen card should be the only card facing up or facing down.

Can you explain why this trick works in terms of the XOR operator? Did you use XOR to calculate that the wining move in the above Nim game? Leave 2 counters in the row with 9. XOR can be used to generate a cellular automata to the Sierpinski triangle, which in turn can be generated from a binary Pascal's triangle using XOR. Pascal's triangle can also provide an algorithm for solving the Towers of Hanoi. Here is a chart showing the various puzzle solved, or generated, by XOR.[5]

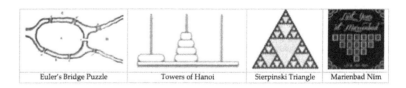

| Euler's Bridge Puzzle | Towers of Hanoi | Sierpinski Triangle | Marienbad Nim |

XOR can be generalized to an infinite-valued logic generalizes the Paradox of the Liar to something called the "Chaotic Liar"–which has the property of begin chaotic in a precisely defined mathematical sense. A dualist version of the Chaotic Liar in fact generates fractal image on the cover of this book!

Notes

Preface

[1] The classic black and white Alice illustrations used in this text are John Tennial's in Lewis Carroll's *Alice's Adventures in Wonderland* (1865) and *Through the Looking Glass and What Alice Found There* (1871).

[2] Gardner, M. (1967), pp. ix-x.

1. The Thinking Reed

[1] Russell. The last sentence of Chapter I, "Appearance and Reality" in *Problems of Philosophy*

[2] Matthews, p. 6.

[3] Adams, p. ix.

[4] Einstein. "Letter to Jacques Hadamard."

[5] Gazzaniga.

[6] Vigneau et al.

[7] Gombrich, pp. 370-1.

[8] Turing, p. 442.

[9] Pascal's *Pensées*, no. 200 Penguin/Krailsheimer translation, no. 347, Brunschvicg edition.

[10] Schor, p. 255. From Rodin's Letter to art critic Marcel Adam in daily newspaper, *Gil Blas*, 1904.

[11] Sagan, p. 210.

[12] Searle, pp. 29-30.

[13] Penrose (1989), pp. 92-5.

[14] Penrose (1994), pp. 151-9.

[15] Franzén, pp. 124-5.

[16] See Mar's review of Franzén's book.

[17] Lucas, p. 112.

[18] Hofstader, p. 697.

[19] Gödel, p. 310.

2. "Eureka!" Problem Solving Heuristics

[1] Newell and Simon, p. 72.

[2] Duncker, pp. 1-2.

[3] Gleick, p. 65.

[4] Adams, pp. 30-1.

[5] Perkins, p. 138.

[6] Bono, p. 13.

[7] Parville.

[8] Albert B. Lord on schemata, as discussed in Perkins, pp. 172-4.

[9] Yuasa.

[10] Higginson and Harter.

[11] Weber.

[12] Poincaré, pp. 14-17.
[13] Beasley, p. 102.
[14] Lakoff and Johnson.
[15] Gick and Holyoak.
[16] Weisberg and Alba.
[17] Lung and Dominowski.
[18] Jen Christensen. "US life expectancy is still on the decline. Here's why."*CNN*, 26 November 2019.
cnn.com/2019/11/26/health/us-life-expectancy-decline-study
[19] Aczel, p. 64.
[20] Peter Winkler, *MoMath Gala*, 21 October 2020.

3. Bridges to Problem Solving

[1] Stein, p. 97.
[2] Gardner, M. (1959), pp. 55-62.
[3] Eames and Redheffer.
[4] Ramsey's life is the subject of a biography by Cheryl Misak, *Frank Ramsey: A Sheer Excess of Powers*.
[5] From the Preface of Ramsey's *The Foundations of Mathematics*, p. vii.

4. Puzzles, Paradoxes & Previews

[1] Mar, Manyakina, and Caffary.
[2] Lonergan, p. 657.
[3] Johnson-Laird and Wason.
[4] Doyle, Part I, Ch. 2.
[5] Thomson, pp. 48-9.
[6] Glymour and Stalker.
[7] ibid, p. 98.
[8] ibid.
[9] ibid.
[10] ibid.
[11] Bickel et al.
[12] Gardner, M. (1974), p. 102.
[13] Wasserstrom, p. 166.

5. Computational Magic

[1] From Alan Turing's [1938] Ph.D. thesis, "Systems of Logic Based on Ordinals."
[2] Conway, p. 266.
[3] See María Manzano's article for an amusing anecdote about Church and computers.
[4] This quote is commonly attributed to David but I have not been able to track it down in Constance Reid's classic biography, *Hilbert*.
[5] For a discussion of the middle-thirds card trick (Chapter 5, exercise #3, p. 155) and its connection with the Sierpinski fractal, see Donald C. Benson, pp. 114 - 21.
[6] "The Dating Game" was adapted from a card trick discussed in Paul Curzon and Peter W. McOwan's book, *The Power of Computational Thinking*....

Our version has two individuals enacting the algorithm simultaneously to keep it more interesting. Both versions are based on an algorithm created by magician and computer programmer Alex Elmsley. In magic circles, Emsley is known for the "Ghost Count" or "Emsley Count" which depends on manual dexterity. Fortunately, this automated card trick depends only on following an algorithm.

[7] See Martin Gardner, "Finger Arithmetic," Ch. 8 in *Mathematical Magic Show* (1965).

6. Cultivating Creativity

[1] As quoted in *Madame Curie: A Biography* (1937) by Eve Curie Labouisse, Part 2, p. 116.
[2] Newell and Simon.
[3] Weizenbaum (1976), pp. 188-9.
[4] Weizenbaum (1966), pp. 42-3.
[5] Miller.
[6] Native American History for Kids. (n.d.). Retrieved from ducksters.com/history/native_americans.php
[7] Scheffler, p. 4.
[8] See Singh.

This mathematical joke in the backdrop of the Simpsons episode, of course, is that the identity appears to be a counterexample to Fermat's conjecture. If you simply type the numbers into a spreadsheet or calculator it appears to be verified because of round-off errors. Any counterexample would disprove Fermat's conjecture solving a problem that had stumped mathematicians for centuries. Fermat's conjecture also has an interesting logical property: if it were to be shown to be undecidable (in the sense of Gödel) it would be true! The reason is that if Fermat's conjecture had a numerical counterexample, it would be refutable by a calculation and so would be decidable.

[9] Mar (1993).
[10] Bellah et al., p. 84.
[11] Goleman, Kaufman, and Ray, p. 172.
[12] Suzuki.
[13] Day, p xl-xli.
[14] Gleick (1993), p. 29.
[15] Goodstein, Feynman, and Goodstein, p. 52.
[16] Feynman.
[17] Perkins, pp. 288-9.
[18] Michon.
[19] Carmichael, Hogan, and Walter, p. 75.
[20] Robson.

Preview

[1] Gardner, M. (1959), pp. 30-2. *Scientific American* received the letter from Willison Crichton and Donald Lamphier of Ann Arbor, MI.

Epilogue

[1] Wang, 9.4.20, p. 317.

Acknowledgements

This book would not have been possible without the help of many colleagues, students, and friends over many years. The seeds were planted when I was a graduate student and given one week to prepare and teach a course on critical thinking at UCLA to over three hundred students. Without help from colleagues and friends, who were master teachers, I would not have survived that ordeal. By a fluke and the enthusiasm of friends, soon there were over a thousand students on the waiting list. The material and manuscript evolved over the years and through connatural selection–with the more effective tools for imparting critical thinking skills and habits surviving the test of time. The manuscript bears the stamp of these lowly origins, which at times are illuminated by the insights of students and colleagues, of which there are too many to mention. I would be remiss, however, if I didn't mention a few friends–Nick Fabry, James Hanink, Donald Kalish, David Kay, William Puett, Paul St. Denis, Jerome Tauber, JoAnne Young.

To all my teachers and students without whom the critically examined life would not be worth living.

To Rosemary Ashton, my wife, without whom this project would not have been undertaken at this point in my life.

Bibliography

Amir Aczel. *Chance: A Guide to Gambling, Love, the Stock Market and Just About Everything Else*. MIT Press, 2004.

James Adams. *Conceptual Blockbusting, 3rd Edition*. W.W. Norton & Company, 1986.

Edward H. Adelson. "Checkers Shadow Illusion." *MIT Website*, 1995. http://persci.mit.edu/gallery/checkershadow.

John Beasley. *The Mathematics of Games*. Oxford University Press, 1989.

Robert N. Bellah et al. *Habits of the Heart: Individualism and Commitment in American Life*. University of California Press, 1985.

Donald C. Benson. *The Moment of Proof: Mathematical Epiphanies*. Oxford University Press, 1999.

E. W. Beth. *The Foundations of Mathematics* (rev. edition.). Harper & Row, 1966.

Peter J. Bickel et al. "Sex bias in graduate admissions: Data from Berkeley." *Science*, vol. 187, no. 4175, 1975, pp. 398–404.

Edward de Bono. *Lateral Thinking: Creativity Step by Step*. Harper and Row, 1970.

Lewis Carroll. *Alice's Adventures in Wonderland* Macmillan, 1865.

————. *Through the Looking Glass and What Alice Found There*. Macmillan, 1871.

Leonard Carmichael, H.P. Hogan, and A.A. Walter. "An experimental study of the effect of language on the reproduction of visually perceived form." *Journal of Experimental Psychology*, vol. 15, no. 1, 1932, pp. 73-86.

John H. Conway and Richard K. Guy. *The Book of Numbers*. Springer-Verlag, 1996.

Alan Cowell. "Overlooked No More: Alan Turing, Condemned Code Breaker and Computer Visionary." *The New York Times*, June 10, 2019. nytimes.com/2019/06/05/obituaries/alan-turing-overlooked.html.

Paul Curson and Peter McOwan. *The Power of Computational Thinking: Games, Magic and Puzzles to Help You Become a Computational Thinker*. World Scientific Publishing Europe, 2017.

Dorothy Day. *Dorothy Day, Selected Writings: By Little and by Little*. Robert Ellsberg (ed.). Orbis Books, 1992 (Knopf, 1983).

Arthur Conan Doyle. *A Study in Scarlet and the Sign of Four*. Smith, Elder & Company, 1903.

Karl Duncker. "On Problem Solving." *Psychological Monographs*, vol. 58, no. 5, 1945.

Charles Eames and Ray Redheffer. Poster/mural, *Men of Modern Mathematics: A History of Mathematicians from 1000 to 1900*, commissioned for IBM Corporation, 1966.

Betty Edwards. *Drawing on the Right Side of the Brain*. J.P. Tarcher, Inc., 1979.

Einstein, Albert. "Letter to Jacques Hadamard." In *The Creative Process: A Symposium*. Brewster Ghiselin (ed.). New American Library, 1952.

Leonhard Euler. "The Königsberg Bridges." In *Mathematics: An Introduction to its Spirit and Use: Readings from 'Scientific American.'* Morris Kline (ed.). W.H. Freeman and Company, 1979.

Richard Feynman. *The Character of Physical Law*. MIT Press, 1967 (1965).

Torkel Franzén. *Gödel's Theorem: An Incomplete Guide to Its Use and Abuse*. A K. Peters, 2005.

Gottlob Frege. "The thought: A logical inquiry." *Mind*, vol. 65, no. 259, 1956 (1918), pp. 287-311.

Robert Frost. "The Figure a Poem Makes." In *Collected Poems of Robert Frost*. Henry Holt and Company, 1939.

Howard Gardner. *Frames of Mind: A Theory of Multiple Intelligences*. Basic Books, 1983.

_____. *Intelligence Reframed: Multiple Intelligences for the 21st Century*. Basic Books, 1999.

Martin Gardner. *Martin Gardner's Science Magic: Tricks & Puzzles*. Dover, 2011. (Sterling, 1997.)

_____. "Mathematical Games." *Scientific American*, March 1974.

_____. *Mathematical Magic Show*. Vintage, 1965.

_____. *The Scientific American Book of Mathematical Puzzles & Diversions*. Simon and Schuster, 1959.

Michael S. Gazzaniga. "The Split Brain Revisited." *Scientific American*, vol. 279, no. 1, 1998, pp. 50–55. *JSTOR*, jstor.org/stable/26057845

Brewster Ghiselin (ed.). *The Creative Process: A Symposium*. New American Library, 1952.

Mary L. Gick and Keith J. Holyoak. "Schema induction and analogical transfer." *Cognitive Psychology*, vol. 15, 1983, pp. 1-38.

James Gleick. *Chaos: Making a New Science*. Viking Penguin Books, 1987.

_____. *Genius: The Life and Science of Richard Feynman*. Vintage, 1993.

Clark Glymour and Douglas Stalker. "Winning through pseudoscience." In *Philosophy of Science and the Occult*, 2nd Edition. Patrick Grim (ed.). SUNY Press, 1990, pp. 92-103.

Kurt Gödel. *Collected Works: Volume III: Unpublished Essays and Lectures*. Solomon Feferman et al (eds.). Oxford University Press, 1986.

Daniel Goleman, Paul Kaufman, and Michael Ray. *The Creative Spirit: Companion to the PBS Television Series*. Dutton, 1992.

Ernst H. Gombrich. *Art and Illusion: A Study in the Psychology of Pictorial Representation.* Princeton University Press, 1961, pp. 370-371.

David L. Goodstein, Richard P. Feynman, and Judith R. Goodstein. *Feynman's Lost Lecture: The Motion of Planets Around the Sun.* WW Norton & Company, 1996.

Patrick Grim (ed.). *Philosophy of Science and the Occult*, 2nd Edition. SUNY Press, 1990.

Patrick Grim, Gary Mar, and Paul St. Dennis. *The Philosophical Computer.* MIT Press, 1998.

Güven Güzeldere and Stefano Franchi. "Dialogues with colorful 'personalities' of early AI." *Stanford Humanities Review*, vol. 4, no. 3, 1995, pp. 161-169.

William Higginson and Penny Harter. *The Haiku Handbook.* Kodansha International, 1985.

Douglas R. Hofstader. *Gödel, Escher, Bach: An Eternal Golden Braid.* Basic Books, 1979.

Philip N. Johnson-Laird and Peter C. Wason. "A theoretical analysis of insight into a reasoning task." *Cognitive Psychology* 1, no. 2, 1970, pp. 134-148.

Flavius Josephus. *Antiquities of the Jews*, c. 93–94 AD.

Nobuyuki Kayahara. "The Spinning Dancer." *Procreo Flash Design Laboratory*, 2003. commons.wikimedia.org/wiki/File:Spinning_Dancer.gif

Helen Keller. *Three Days to See*, 1933.

John F. Kennedy. "1959 Convocation of the United Negro College Fund." Indianapolis, Indiana, 12 April 1959. jfklibrary.org/asset-viewer/archives/JFKCAMP1960/1029/JFKCAMP1960-1029-036.

Morris Kline (ed.).*Mathematics and the Modern World: Readings from 'Scientific American.'* W.H. Freeman and Company, 1968.

_____. *Mathematics: An Introduction to its Spirit and Use: Readings from 'Scientific American'.* W.H. Freeman and Company, 1979.

Gina Kolata. "At last, shout of 'Eureka!' in age-old math mystery." *New York Times*, Section A, p. 1, 24 June 1993. nytimes.com/1993/06/24/us/at-last-shout-of-eureka-in-age-old-math-mystery.html.

_____. "How a gap in the Fermat Proof was bridged." *New York Times*, Section C, p. 1, 31 January 1995. nytimes.com/1995/01/31/science/how-a-gap-in-the-fermat-proof-was-bridged.html

Saul A. Kripke. *Wittgenstein on Rules and Private Language: An Elementary Exposition.* Harvard University Press, 1982.

Thomas Kuhn. *The Structure of Scientific Revolutions.* University of Chicago Press, 1962 (1970, 1996).

George Lakoff and Mark Johnson. *Metaphors We Live By.* University of Chicago Press, 1980.

David Lewis and James Greene. *Thinking Better*. Holt, Rinehart, and Winston, 1983 (1982).

Bernard Lonergan. *Insight: A Study of Human Understanding*, Vol. 3. University of Toronto Press, 1992.

Albert B. Lord. *The Singer of Tales*. Harvard University Press, 1960.

J.R. Lucas. "Minds, Machines and Gödel." *Philosophy*, 1961, pp. 112-27.

Ching-Tung Lung and Roger L. Dominowski. "Effects of strategy instructions and practice on nine-dot problem solving." *Journal of Experimental Psychology: Learning, Memory, and Cognition*, vol. 11, no. 4, 1985, pp. 804-811.

María Manzano. "Alonzo Church: His life, his work and some of his miracles." *History and Philosophy of Logic*, vol. 18, no. 4, 1997, pp. 211-32.

Gary Mar. "Euthanasia: A Personal Perspective." *Contexts: A Forum for the Medical Humanities*, vol. 2, no. 2, Dec. 1993.

_____. "Hao Wang's Logical Journey." *Journal of Chinese Philosophy* vol. 42, 2015, pp. 540-561.

_____. Review of Franzén's *Gödel's Theorem: An Incomplete Guide to Its Use and Abuse*. In *The Mathematical Intelligencer*, vol. 29, no. 2, 2007, pp. 66-70.

Gary Mar, Yuliya Manyakina, and Amanda Caffary. "Unless and until: A compositional analysis." In *International Tbilisi Symposium on Logic, Language, and Computation*. Springer, Berlin, Heidelberg, 2013, pp. 190-209.

Gareth Matthews. *Philosophy and the Young Child*. Harvard University Press, 1980.

John A. Michon. "The game of JAM: An isomorph of Tic-Tac-Toe." *The American Journal of Psychology*, vol. 80, no. 1, 1967, pp. 37-140.

George Miller. "The magical number seven, plus or minus two: Some limits on our capacity for processing information." *Psychological Review*, vol. 63, 1956, pp. 81-97.

Cheryl Misak. *Frank Ramsey: A Sheer Excess of Powers*. Oxford University Press, 2020.

Roice Nelson. "Abstracting the Rubik's Cube." In *The Best Writing on Mathematics 2019*. Mircea Pitici (ed.). Princeton University Press, 2019, pp. 43-52.

Allen Newell and Herbert Simon. *Human Problem Solving*. Prentice-Hall, 1972.

Matt Parker. *Things to Make and Do in the Fourth Dimension*. Farrar Straus and Giroux, 2014.

Henri de Parville. *La Nature* (translated by W.W.R. Ball.)

Blaise Pascal. *Pensées*.

John Allen Paulos. *Innumeracy: Mathematical Illiteracy and Its Consequences*. Hill and Wang, 1989, 2001.

Roger Penrose. *The Emperor's New Mind*. Oxford University Press, 1989.

———. *Shadows of the Mind*. Oxford University Press, 1994.

David N. Perkins. *The Mind's Best Work*. Harvard University Press, 1981.

Mircea Pitici. *The Best Writing on Mathematics 2019*. Mircea Pitici (ed.). Princeton University Press, 2019.

Henri Poincaré. "Mathematical Creation." In *Mathematics and the Modern World* Mathematics in the modern world: Readings from 'Scientific American.' Morris Kline (ed.). W.H. Freeman and Company, 1968.

George Polyá. *How to Solve It*. Princeton University Press, 1945.

Frank Plumpton Ramsey. *The Foundations of Mathematics*. Littlefield, Adams & Company, 1960.

Constance Reid. *Hilbert*. Springer, 1970.

David Robson. "A new trial of an ancient rhetorical trick finds it can make you wiser." *Research Digest*, 10 July 2019. digest.bps.org.uk/2019/05/24/a-new-trial-of-an-ancient-rhetorical-trick-finds-it-can-make-you-wiser/.

Bertrand Russell. *The Problems of Philosophy*. Henry Holt and Company, 1912.

Carl Sagan. "The fine art of baloney detection."*The Demon Haunted World*. Random House, 1995.

Israel Scheffler. "In Praise of the Cognitive Emotions." *Routledge*, 1991 (1976).

Naomi Schor. "Pensive texts and thinking statues: Balzac with Rodin." *Critical Inquiry*, vol. 27, no. 2, 2001, pp. 239–265. *JSTOR*, jstor.org/stable/1344249.

Kathryn Schultz. *Being Wrong: Adventures in the Margin of Error.* Ecco, 2010.

John Searle. *Minds, Brains, and Science*. Harvard University Press, 1990.

Simon Singh. *The Simpsons and Their Mathematical Secrets*. Bloomsbury USA, 2013.

Roger W. Sperry. "Hemisphere disconnection and unity in conscious awareness." *American Psychologist*, vol. 23, 1968, pp. 723-33.

Sherman K. Stein. *Mathematics: The Man-made Universe*. W.H. Freeman and Company, 1969.

Ian Stewart. "A partly true story." *Scientific American*, vol. 268, no. 2, February 1993, pp. 110-112.

Shinichi Suzuki. *The Writings of Shin'ichi Suzuki: Where Love is Deep*. Talent Education Journal, 1982.

Allen Tager. "Why was the color violet rarely used by artists before the 1860s?: A descriptive summary and potential explanation." *Journal of Cognition and Culture*, vol, 18, nos. 3-4, 13 August 2018. doi.org/10.1163/15685373-12340030

Judith Jarvis Thomson. "A defense of abortion." *Philosophy and Public Affairs*, vol. 1, no. 1, pp. 47-66, 1971.

Alan Turing. "Computing machinery and intelligence." *Mind*, vol. 59, no. 236, 1950, pp. 433-460.

Mathieu Vigneau et al. "What is right-hemisphere contribution to phonological, lexico-semantic, and sentence processing?: Insights from a meta-analysis." *Neuroimage* vol. 54 no. 1, 2011, pp. 577-93.

Hao Wang. *A Logical Journey: From Gödel to Philosophy*. MIT Press, 1996.

Richard Wasserstrom. "The University case for preferential treatment." *American Philosophical Quarterly*, vol. 13, 1976, pp. 165-170.

Robert Weber. *A Random Walk Through Science*. Institute of Physics, 1973.

Robert W. Weisberg and Joseph W. Alba. "An examination of the alleged role of 'fixation' in the solution of several 'insight' problems." *Journal of Experimental Psychology: General*, vol. 110, 1981, pp. 169-192.

Joseph Weizenbaum. *Computer Power and Human Reason: From Judgment to Calculation*. W. H. Freeman and Company, 1976.

_____. "ELIZA - A computer program for the study of natural language communication between man and machine." *Communications of the ACM (Association for Computing Machinery)*, vol. 9, no. 1, 1966, pp. 36-45.

Eugene Wigner. "The unreasonable effectiveness of mathematics in the natural sciences." *Communications on Pure and Applied Mathematics*, vol. 13, 1960, pp.1-14.

Andrew Wiles. "On solving Fermat." *Nova* website, 1 November 2000. pbs.org/wgbh/nova/article/andrew-wiles-fermat/

Ludwig Wittgenstein. "Philosophical Investigations." *Philosophische Untersuchungen*, 1953.

Nobuyuki Yuasa. *Basho, The Narrow Road to the Deep North and Other Travel Sketches*. Penguin Books, 1966.

Index

Adelson, Edward H., 171
Alba, Joseph W., 49
algorithm, 137, 150, 153
analogy, 103, 105
Archimedes, 39
argument, 84–86
   persuasive, 102
   rhetorical, 103
Aristotelian categorical propositions, 99
Aristotelian Square of Opposition, 97
artificial intelligence, 6, 16, 160

Basho, 34
biconditional, 77
binary coding, 141, 150
brain, left-right, 4, 5, 9
brainstorming, 26
Bridges of Königsberg, 52
   model, 57, 62
Brown, Gordon, 149

Cantor's discontinuum, 152
card tricks, 138, 151
Carmichael, L., 184
categorical proposition, 97
chaos theory, 31
checker-shadow illusion, 171
Church's Theorem, 148
Church, Alonzo, 144, 148
Clark, Arthur C., 151
complex sentence, 76, 78
complexity, 31
computational paradox, 160

computational thinking, 137, 150
conditional, 77
conjunction, 76
connected graph, 52
convergence schema, 48
Conway, John Horton, 144
correspondence, 196

data architecture, 139, 150
Day, Dorothy, 177
debugging, 150
dependent relations, 94, 117
Descartes, René, 169
Diaconis, Persi, viii
dictionary, 77
disjunction, 76
Dodgson, Charles (Lewis Carroll), 83
Dominowski, Roger L., 49
Dudeney, Henry Ernest, 65

Einstein, Albert, 4
   thought experiments, 31
Elmsley, Alex, 153
emotions, 165
Enigma code, 149
Euler path, 53
Euler's theorem, 55, 60
Euler, Leonhard, viii, 52, 63
exclusive 'or', 201
expert knowledge, 165
expert systems, 161

fallacy, 85, 86
   false analogy, 106
   faulty generalization, 107

faulty precedent, 106
Fermat's Last Theorem, 169
Fermat, Pierre, 169
Feynman, Richard, 152
formal language, 196
fractals, 31
Franzén, Torkel, 16
Frege, Gottlob, 195
Fuller, Buckminster, 57

Gardner, Howard, 184
Gardner, Martin, viii
   letter from reader, 199
generalization, 105
geodesic dome, 57
Glymour, Clark, 108
grammatical tree, 78
Gödel, Kurt, 148
   with Hao Wang, 201
Gödel's theorem, 17

haiku, 34
Halmos, Paul, 77
halting problem, 148
Hamilton, William R., 59
Hamiltonian path, 59
heuristics, 22, 39, 162
hierarchical knowledge
   networks, 165
Hilbert, David, 148
Hofstader, Douglas, 16
Human Potential Movement, 3

illusion, 171
Imitation Game (film), 149
incompleteness theorem, 148
inference rules, 84
invalid argument, 86
isomorphism, 61, 166

Josephus problem, 143
Josephus, Titus Flavius, 143

Kant, Immanuel, 60
Kayahara, Nobuyuki, 172
Kripke, Saul, 187
Kuratowski's theorem, 67

Ladd-Franklin rules, 101
Ladd-Franklin, Christine, 101
lambda-calculus, 148
laws of truth, 196
Leibniz, Gottfried, viii, 52,
   156, 196
Lewis Carroll, 83, 188
liar, 197
   chaotic liar, 202
Lucas, J.R., 16
Lucas, Édouard, 58
Lung, Ching-tung, 49

matrix, 117
matrix logic, 69, 73
memory
   chunking, 162
memory palace, 163
mentalism, 151
Michon, John A., 183
modeling, 56
modus ponens, 84, 167
modus tollens, 84
moral permissibility, 104

natural language, 167
New York Times
   Alan Turing obituary, 148
Newman, Max, 148
Nim, game of, 201
nine dot puzzle, 23, 49

paradox
   Newcomb's, 114
   Russell's, 195
   Simpson's, 110
parity, 54, 202

peg-word coding, 163
Penrose, Roger, 16
Perkins, D.N., 181
Poincaré, Henri, 31
    mathematical creativity, 38
    three body problem, 31
polarity, 5
possible worlds, 196
precedent case, 105
probability, 112
probability tree, 114, 119
proof, 146
pseudoscience, 107
puzzle, 57
    Angels & Demons, 199
    Euler's Pond, 60
    Fork in the Road, 195
    Kant's Ghost, 61
    Towers of Hanoi, 32, 58, 160

Ramsey's theorem, 68
Ramsey, Frank P., 68
reductio ad absurdum, 28
reverse engineering, 150
rhetoric, 102
Ricci, Matteo, 163
Rogers, Carl Ransom, 161
Russell, Bertrand, 1, 195

Sagan, Carl, 13
Scheffler, Israel, 165
schemata, 33
Searle, John, 16
sequencing, 91
Sherlock Holmes, 90
Sierpinski triangle, 152
Sierpinski, Waclaw, 144

silhouette illusion, 172
Simonton, Dean, 165
spinning dancer, 172
Stalker, Douglas, 108
strong AI, 17
subalternation, 98
subgoaling, 32
syllogism
    reductio set, 99

Thomson, Judith Jarvis, 103
    abortion debate, 103
transitive inference problem, 6, 12
truth, 196
truth table, 75, 125, 197
truth-functionality, 76
Turing, Alan M., 6
    Entscheidungsproblem (Decision Problem), 148

unless debate, 73, 82

valid argument, 86
Venn diagram, 98

Watergate Impeachment Proceedings, 73
Weisberg, Robert W., 49
Weizenbaum, Joseph, 161
Wiles proof, 170
Wiles, Andrew, 169
Wittgenstein, Ludwig, 187
World War II, 149

XOR (exclusive 'or'), 201

CPSIA information can be obtained
at www.ICGtesting.com
Printed in the USA
BVHW051220130821
613572BV00001B/1